POWER STRUGGLES

NEW ANTHROPOLOGIES OF EUROPE

Michael Herzfeld, Melissa L. Caldwell, and
Deborah Reed-Danahay, *Editors*

POWER STRUGGLES

Dignity, Value, and the Renewable Energy Frontier in Spain

Jaume Franquesa

Indiana University Press

This book is a publication of

Indiana University Press
Office of Scholarly Publishing
Herman B Wells Library 350
1320 East 10th Street
Bloomington, Indiana 47405 USA

iupress.indiana.edu

The paper used in this publication meets the minimum requirements of
the American National Standard for Information Sciences—Permanence
of Paper for Printed Library Materials, ANSI Z39.48-1992.

Manufactured in the United States of America

Cataloging information is available from the Library of Congress.

ISBN 978-0-253-03373-4 (cloth)
ISBN 978-0-253-03372-7 (paperback)
ISBN 978-0-253-03376-5 (ebook)

1 2 3 4 5   23 22 21 20 19 18

Per na Marion.

Contents

Illustrations follow page 100.

Acknowledgments

THIS BOOK WOULD not have become a reality without the help of many people, and I am grateful to all of them. But first things first, for only the hospitality of the citizens from Southern Catalonia, and especially of those from Fatarella (*una dura terra que estimo*, as the poet wrote), made my research possible. Unfortunately, I cannot name them all. My special thanks go to Joan de Marta and Sisco de Permarch, with their families and *colla* (Encarna, Dolors, Ramon, Natàlia, Andreu, Maria, Vinyes, Miquel Àngel, etc.), Joan de l'Àngela and his family, Josep i Núria de Sagalo, Gemma i Josep de Pastoret, Rosa Ruiz, Pol and Mercè, Marta and Gatano de Gironès, Messe Cabús, Sisco de Belart, Ruth and Marco, Josep Maria and Rosa, Ramona i Ramon de Segura, Cinta and Sisco de Carlets, Andreu, Hèctor, Raül, and last, but not least, Montse, Rosa, and Carme, the three wonderful teachers of the *guarde*.

Local institutional support was also key. During my fieldwork, Fatarella had three different mayors (Carme, Fermín, and Jesús) from three different political parties: all of them supported my research and allowed me access to the municipal archives; former mayors Blanch, Basco, and Suñé also provided help and insight. Several mayors and former mayors from other Southern Catalan towns shared their perspectives with me, including the Mesa d'Alcaldes per al Desenvolupament de l'Energia. I also benefited from the experience and wisdom of the Southern Platforms, especially those of Priorat and Terra Alta; the help offered by Roser Vernet, Jordi Clua, and Txus Carbó deserve special mention. Above all, the people of Fundació el Solà, with Neus Borrell and Josep Maria Font at the fore, provided an invaluable intellectual atmosphere: their work is a model, and although they may not know it, they gave me the idea for this book. I also appreciate the willingness of several members of EolicCat (the Catalan association of wind energy producers) and other people in the wind energy sector to meet with me on repeated occasions and share their point of view; my admiration goes to those who believe in building a *different* tomorrow.

A Hunt Fellowship from the Wenner Gren Foundation (Gr. 8732) provided crucial financial support and a good deal of emotional relief for the writing of this book. My own institution, the University at Buffalo–SUNY has provided important support at different phases of this project, fundamentally through two research grants from the Baldy Center for Law & Social Policy, the Julian Park

Publication Fund and a fellowship from the Humanities Institute that permitted me to complete the manuscript. I also want to thank the Universitat Rovira i Virgili, where I enjoyed a short visiting Fellowship in 2010. Between 2012 and 2015 I was involved in the research project "Addressing the Multiple Aspects of Sustainability," based in the Universitat de Barcelona and directed by Susana Narotzky, who, as always, has provided crucial support and guidance for this book.

I cannot express how fortunate I was for receiving the week-long visit of Gavin Smith while I was on the field; from him I have learnt how to be a better researcher, a better thinker, and a better person; many of the ideas in this manuscript have their origin in conversations that we started in Fatarella and continued in Toronto, where he and Winnie Lem have provided an academic home away from home. Marc Morell and Jaime Palomera also visited me in the field: we had a blast, mixing friendship and ideas. Several Southern Catalan academics and intellectuals—Xavier Garcia, Josep Sánchez Cervelló, Jordi Ferrús, Joan Asens, Sergi Saladié, and Joan Rebull—greatly helped me refine my thoughts and perspective while on the field.

Several people read drafts of parts of the manuscript, always giving me useful feedback: the late Susan Christopherson, Blanca Garcés, Vasiliki Neofotistos, Susana Narotzky, Pablo Sánchez León, Gavin Smith, Marion Werner, and Saulesh Yessenova. Drew Gilbert, Irene Ketonen, Tania Li, Deborah Reed-Danahay, Beatriz Santamarina, and Tom Wilson gave me the opportunity to present my work in front of small and larger audiences. Many other scholars contributed their insights, probably unknowingly, through formal and informal discussion: Ayse Caglar, Sandra Ezquerra, Bilge Firat, Víctor Giménez, Kregg Hetherington, Don Kalb, Aaron Kappeler, Joan Martínez Alier, Carlotta McAllister, Patrick Neveling, Don Nonini, Florin Poenaru, Scott Prudham, Katharine Rankin, Guillermo Sanz, and Sandy Smith-Nonini. My colleagues and graduate students at the University at Buffalo also have my gratitude for providing a collegial atmosphere. Enze Zhang helped me with the maps, Letta Page with editing, and Ariel Noffke with the bibliography. I am also grateful to Indiana University Press for believing in this project.

This book has proven to be a more challenging task than I had anticipated when I began working on it in 2010. It all happened at once: getting used to a new academic context and a new country, writing in a foreign language, learning to be a parent, engaging with a new field site, and a new set of theoretical questions. The list could go on, and on many occasions, I thought I had overextended myself. Sometimes acedia ensued, and all too often I was too distant from those whom I love. I shall thus extend my deepest gratitude to them for coping with me: my friends in Igualada, Barcelona, Palma, Toronto, Buffalo, and elsewhere; my parents Dolors and Jaume, my siblings Oriol and Marcel·la (and her beautiful family), and my grandparents, Josefina and Jaume; my Canadian family, Beatrice and Daniel, for giving me the time, the trust, and the space. Yet if there is one person to whom I owe it all, it is Marion Werner: for her love, her courage, her intelligence, her sense of purpose, and, not least, for her intellectual stimulus and her help with the manuscript. And finally, to our children, Yannick and Biel, a constant source of happiness: *perquè voldreu.*

POWER STRUGGLES

Where the World Ends

Doesn't a breath of the air that pervaded earlier days caress us as well?
—Walter Benjamin, *Theses on the Philosophy of History*

"Hey, you, TORONTO, what are you doing on Sunday?" Bernat asked the question without raising his head. He was focused on his job: carrying a load of just-harvested almonds from his small tractor and laying them to dry on a tarp outside his house. Rain was the least of his worries: four long months had gone by since the last rain, back in May.

Toronto—the city where I was living when I began this research—had become my nickname among the villagers of Fatarella, in the Southern Catalan county of Terra Alta (High Land). "Everyone has a name here, otherwise no one would know who you are," Bernat had once told me. Where last names are for outsiders and official institutions and first names are confined to immediate family, nicknames authenticate the visitor.

Bernat was my neighbor. His principal nickname was also the name of his household, an old and respected one in the village; the year 1700, sculpted in the lintel of his house, quietly asserted that temporal depth. Among the first people I met in Fatarella, he was completing a two-year term as president of the local agrarian cooperative in 2010. Bernat was a *pagès* (peasant, farmer), but despite owning a fair amount of land, he had to supplement his agrarian income with temporary jobs in the nuclear plants of Ascó, just ten kilometers downhill from the village. His wife worked as a cashier in a supermarket of Móra, the main commercial town of the area, and his daughter attended high school in Gandesa, the county capital.

"I have no plans. Any suggestions?" I replied.

"We are going on an excursion. You can join us, but be ready to walk." He added, "It'll be the whole *colla*." That is, Bernat would be joined by a healthy quorum of his long-standing group of friends—male and female—formed during adolescence.

"Great, where are we going?"

"Where the world ends," he answered with shiny eyes. Then he paused and switched the topic. "Do you have almonds? No, you don't, my lord, you have nothing. Come, let's go inside; you need a bag."

We entered the ground floor of his house, a threshold between the domestic space, occupying the second floor, and the rest of the village, extending through the fields, collectively called *defora* (literally, "outside"). Tractors have long replaced the donkeys and pigs that used to inhabit the basements, but the deposits for the grapes and the shelves with preserves are still there, along with a busy collection of tools and heteroclite objects.

Plastic bags were kept in an old ammunition box, a remnant of the bloody Civil War Battle of the Ebro, which had unfolded in Southern Catalonia in 1938: "I found it a few years ago while picking mushrooms. Isn't it cool? We don't throw anything away here." The war was far from a mere relic, though. Bernat's own grandfather had been killed by an anarchist co-villager during the war and Bernat's late father had never accepted his son's left-wing sympathies, considered a betrayal to the memory of that dark time. Bernat recalls his father's reaction to his political awakening: "We worked side by side, every day in the fields, and for at least a year we did not exchange one word."

He gave me a bag so that I might fill it with almonds, and poured wine into a large, empty soda bottle. The liquid gift shone with the typical amber color of the local white Grenache variety. During the nineteenth century, its unusually high alcoholic content had made it a coveted commodity, especially for French merchants who would use it to bulk up their own product.

"I make it myself," said Bernat proudly, "old style, with my feet. But I don't like wine, it's for my in-laws."

Sunday arrived and I met the *colla*, about twelve people, at the huge bar of the *casal*, a municipally owned building that houses all sorts of public services— daycare, library, outdoor swimming pool, ball room, hunters' society, and so forth. The *casal* was built in the 1970s with supplies and labor donated by the villagers. After a coffee, we started walking toward the hilly area known as Valencians, leaving the village behind. The surprisingly wide, uphill path took us through fields filled with almond and hazelnut trees, olive groves, stone walls, and wind turbines. The friends' chatter about the fields gave the excursion the feel of a reconnaissance mission: "Look at these weeds, they're eating the field. Doesn't Catarro own this? I see that his children are not taking care of it."

After a couple of hours hiking, we gathered for a snack of nuts and fruit. Less than a hundred meters from where we were stood an electricity substation connected to the wind farms, yet no one mentioned it. In fact, everyone seemed keen to avoid a topic—the installation of the wind farms—that had opened cleavages within *colles* and families. An electricity line left the station and accompanied our journey until the end of the path (I later learned that the electricity company had widened the path just a few years earlier).

Past noon, the path ended. We had arrived at our destination, a *mas* belonging to Feliu, one of the excursionists. *Masos* are humble dwellings made with *pedra seca* (drystone, or stone construction with no mortar). In the past, they served as winter refuges during the harvest, when the short days and long hours of labor made it impossible to return to the village to sleep, but nowadays they have become spaces of leisure for *colles* and families, modest paradises used to socialize during the weekends. We went to the forested patch beyond the *mas* to admire the myriad ingenious stone traps—a traditional bird-hunting technique—that Feliu's brother had set up. Several excursionists confided: "They are illegal."

We made a fire with dead branches hastily gathered and set the table, grabbing an assortment of mismatched chairs and situating them around a makeshift table of sawhorses and boards. As the group bustled, Bernat approached. In his typical subdued yet cocky way, he whispered: "Do you want to see where the world ends?"

We walked a couple hundred meters to the north, toward the river, as the wind grew ever stronger. Arriving at the edge of a cliff, Bernat opened his arms: "This is the end of the world." Beneath the 300-meter high scarped cliff, the Ebro ran lazily through a wide canyon. The ancient historian Herodotus had called it Iber, and the whole peninsula came to be known as Iberia. It was the western end of the ancient Mediterranean world.

Yet Bernat did not have the ancient Greeks in mind. Nor did he have political borders in mind either, though we were standing close to the limit between Aragon and Catalonia, for people on both sides are knitted together by friendships and marriages, sharing forms of livelihood and the Catalan language. In some way, for Bernat this was the end of the world because it was the end of his agrarian landscape, the outermost confines of Fatarella. The path ended at the edge of the *garriga* (forest), a symbolically threatening space beyond household nicknames and beyond the *cura* (care, management) of the peasant, inhabited by wild animals and the memories of the *maquis* (the anti-Francoist guerrilla). Thus, in the local parlance the expression "to become forest" refers to any human product that has lost direction and cohesion: a field laying waste, an abandoned agrarian landscape, a household losing its property and its name, a village consumed by insidious infighting.

And yet Bernat was not looking to the forest. His eyes were fixed downward on the sobering landscape traversed by the river making its way to the sea, his face beaten by the strong wind that makes his village so attractive to renewable energy developers. The relationships that link the inhabitants of Fatarella and the rest of Terra Alta to the riverine villages of the county of Ribera are even stronger than those extending beyond the Aragonese border. The worlds of those villages are not unlike his. Bernat was seeing something else: a synchronic picture of half a century of energy development.

The sky was clear, affording the eye a privileged, if somehow unsettling view. One could follow the double curve of the Ebro, first westward to the border between Catalonia and Aragon, and then, past Flix and Ascó, southward toward the sea, completing a long trip through dry, rural lands, connecting, as only the Ebro does, Atlantic Iberia and Mediterranean Iberia. At its ends lay the only two historically large Spanish industrial poles, the Basque one around Bilbao, and the Catalan cluster, with Barcelona at its center. The image was daunting: two thirds of the electricity consumed in Catalonia is generated in this small, forgotten corner of the country, and one could *see* it.

At the far west, the hydroelectric dams and reservoirs of Mequinensa and Riba-roja, two of the largest in Spain, offered the image of a human-made inland sea. Past the second dam, the river recovered much of its shape until reaching Flix, where one last, smaller dam provided electricity to an electro-chemical plant, installed by a German company at the end of the nineteenth century to take advantage of the cheap hydro energy. There, decades of chemical waste dumping gave the river an orange tone.

Further down stood the two Faustian chimneys of Ascó's nuclear plants, connected by electric cables to the coal plants located beyond the Aragonese border. Across the river, just above dry fields and a growing *garriga*, three busy wind farms and one solar farm rose up the slopes that connected the counties of Ribera and Priorat. Further down the stream, toward Móra and Xerta, a fertile irrigated plain, filled with peach, orange, and cherry trees, betrayed the eye. These fields owed their existence to a long, local struggle that had prevented the area, first, from being flooded by two more hydroelectric dams, and later, at the turn of the century, from hosting a large natural gas power plant. These were heroic deeds recounted in a local mythology of resistance.

Past the river, in a straight line of just forty kilometers, there was the sea. The eye could not penetrate the coastal mist, but I knew that two more large energy facilities—one using natural gas, the other nuclear power—stood in that blurry, bluish band. Next to them stood a second nuclear plant, partially dismantled after a frightening accident in 1989. Just a few kilometers to the south, the fishermen of Ametlla still remember their struggle to prevent the installation of another two nuclear plants.

I turned my back to the river and looked to the plateau of Terra Alta, where eight wind farms extended north to south in a continuous line. Many more were certainly on the books. Then I looked at the electric line that had accompanied us since we went through the substation, and observed how it turned downhill toward the "end of the world" to reach Riba-roja's hydroelectric power plant. From there, the cables connected Fatarella's wind with the wind on the other bank of the Ebro, with the hydropower from the river, with the atoms being split in Ascó, the Aragonese coal, and the natural gas arriving from Algeria. All these elements bulked together, converted into one indistinguishable abstract potency—energy, electricity—moving as a flow through the abstract space of cables and power plants until reaching the urban metropolitan areas, making possible all sorts of marvels, starting with the turning on of the switch that we, sometimes, estranged, take as the quintessential modern miracle.

"Now you *saw* it," said Bernat. "The world ends here."

"It also starts here," I replied. "This is the engine room of the country."

"No, the world ends here; that is another world. Besides, who wants to look at the engine room?" We went back to the *mas*, retracing our steps, enjoying in anticipation the thick wine, the roasted pig meat, the toasted bread with tomato, the salty herring, the bitter olives, the excessive amount of *allioli*, and the easy conversation, extended with coffee and cigarettes. Three friends of the *colla* would come in cars to pick us up afterward.

This book is about two worlds: about the political economy of energy production and about the livelihoods of Southern Catalans. Throughout its pages, the book offers an ethnographically situated, kaleidoscopic history of Southern Catalonia focused on how its inhabitants engaged with energy projects and facilities throughout the last half-century. My conviction is that this analysis offers a privileged window for the understanding of our energy present and futures.

The Power of Energy

The political economy of energy was strangely on display from the cliff overlooking the Ebro that Sunday. The view from the end of the world left me with a sort of detached awe, triggered by the unusual experience of *seeing* energy—"Now you saw it," said Bernat. Indeed, most people only notice energy when it "malfunctions": a blackout, an oil spill, rising electricity prices, or a contentious pipeline.[1]

The invisibility of energy is truly remarkable: after all, the use of increasing amounts of energy is one of the main markers of the last two centuries, a historical period that Love and Isenhour call "high-energy modernity."[2] During the twentieth century alone, global energy use rose nearly tenfold.[3] Megacities, private automobiles, airports, and mechanized agriculture—it's all been made possible by abundant energy, and, more specifically, through the extensive exploitation of fossil fuels.

High-energy modernity has transformed our experience of the world, giving rise to and sustaining societal arrangements and habits of thought. The belief in the inevitability of progress, as well as its reduction to material growth, are paradigmatic examples, leading to a society in which, in Laura Nader's expression, "economic goals have been substituted for social goals; what are properly means have been elevated to the rank of ends."[4] Timothy Mitchell has argued that these "ways of living and thinking" have one thing in common: "they treat nature as an infinite resource."[5] Because the world is not infinite, he adds, these ways of thinking and living are unsustainable. And yet, our comprehension of this contradiction is burdened by the invisibility of energy as the basis of modernity.

This invisibility is the expression and central attribute of the power of energy. The felicitous polysemy of the word *power* rightly suggests the strong connection between energy and political power: between power as productive force (power over nature) and power as a structure of domination (power over people).[6] Horkheimer and Adorno argued as much in 1947: "What human beings seek from nature is how to use it to dominate it wholly *both it and human beings*. Nothing else counts."[7] Yet, the power of energy has a third, ideological dimension manifested in its capacity to obscure that connection, thus reinforcing political domination and concealing its unsustainable ecological basis.[8]

The invisibility of energy is a key expression of this ideological power. I call it the *ideology of energy* and it is a particular product of high-energy modernity. It consists of conceiving of and presenting energy as an abstract entity, detaching it from actual social relations, making it largely invisible to social analysis. This ideology is consistent with the form of relation with the environment that characterizes capitalist modernity. Anna Tsing, drawing on Marx, calls this form of relation *alienation*:

> Investors ... imbue both people and things with alienation, that is, the ability to stand alone, as if the entanglements of living did not matter. Through alienation, people and things become mobile assets from other life worlds, elsewhere.... Alienation obviates living-space entanglement. The dream of alienation inspires landscape modification in which only one stand-alone asset matters; *everything else becomes weeds or waste*.... When its singular asset can no longer be produced, a place can be abandoned. The timber has been cut; the oil has run out; the plantation soil no longer supports crops. The search for assets resumes elsewhere. Thus, simplification for alienation produces ruins.[9]

Fetishization is the counterpart of alienation; once detached from social relations, things are fetishized, appearing as if they possess agency and productive capacity on their own. Defetishizing energy is both the goal and the point of departure for its anthropological analysis. As Tsing suggests, this defetishization involves bringing into view, and ultimately enabling, the "entanglements" of human and non-human that are at the core of energy systems, thus allowing the analysis of energy to illuminate what Kalb and Tak call the "critical junctions" of social reality.[10]

In Southern Catalonia, these entanglements are complex, for the world of energy generation is at once inseparable from and incommensurable with the experience and moral universe of a Southern Catalan *pagès* such as Bernat. His insistence that the industrial energy and peasant landscapes are two different worlds asserted their incommensurability. Indeed, although coexisting, those two worlds seem to inhabit a different time and space. They are juxtaposed rather than integrated, simultaneous rather than contemporaneous.

These two worlds exist in different spaces and operate at different scales. One is a qualitative world, the texture of everyday life, a universe where every name, every piece of land, every food is signified, pervaded by multiple relations rooted in history. The other one, the world of energy, operates in an abstract space, made of magnitudes and flows, the strategic space of state and capital, connecting nature and power, overcoming all obstacles, insensitive to any particularism.

Although simultaneous, both worlds inhabit nonsynchronous times.[11] In Bernat's peasant world, the past is inscribed in the present, an ongoing praxis to reproduce existing livelihoods into the future. Every future, every project thus emerges as the continuation of an old struggle to bring to life a world where autonomy is possible and dignity recognized. In contrast, in the world of energy, time is the teleological time of progress and power: increased electricity production, more economic growth, and technological advance. This unilineal narrative is often glossed as one of substitution of energy sources: from water to coal, to nuclear power, to natural gas, to, finally, renewables, the ultimate step that some believe will liberate humankind from resource scarcity and pollution.

Separate, yes, but these two worlds are still entirely inseparable. Their histories have been deeply intertwined since at least the 1960s. Southern Catalonia became an energy hub just as it experienced a massive exodus of population, painful manifestation of a process of modernization—mythologized in the Spanish imagination as "the economic Miracle" (see chapter 1)—that affected the whole geography of Spain. Since then, Southern Catalonia has become more and more central to the world of energy production even as it comes to be seen as an increasingly peripheral and impoverished rural region. Agrarian livelihoods here are becoming more and more precarious, and its social order is more and more fragile. As never-ending energy projects make their way into the region, Southern Catalans are largely left out of that modernization. They simply host and built the infrastructure that stands at its base.

Indeed, Southern Catalan livelihoods traverse both worlds. We saw it in Bernat's case, and we will see it elsewhere. Southern Catalans built the nuclear plants, then used their salaries to buy new tractors, fix their houses, buy new land, and also, in fewer cases, abandon agriculture altogether (chapter 3). Their agricultural fields host turbines, cables and the entire related infrastructure that wind farms involve, making possible the much-praised Spanish transition to renewable energy (chapters 6 and 7).

Both worlds are also intimately intertwined in the political trajectory of the region. The nuclear plants triggered a potent local antinuclear movement that catalyzed the desire for political change during the 1970s and 1980s (chapter 2). At the turn of the century, opposition to a series of hydraulic and energy projects snowballed into a massive wave of protest—the Southern Revolt, as it came to be known—that critically eroded the Catalan and Spanish conservative governments (chapter 4). The Southern Revolt was an early, resounding indictment of the socioecological unevenness sustaining the cycle of capital accumulation up until 2008, a period that I call the Second Miracle that was intimately connected to the unparalleled development of wind energy (chapter 5). During all this time, Southern Catalans protested against the role that they were assigned in modern Spain, imagining other energy models.

Southern Catalonia allows us a privileged vision of the world of energy precisely through this contradictory combination of inseparability and incommensurability. From the end of the world, the abstract space of energy looks eerily concrete, its triumphant time awkwardly mundane. The space looks less abstract, not only made up of flows and infrastructure, but also an inhabited space, where winds have names, where nuclear power triggers heavy silences, where water is scarce, where projects are resisted, and where land is owned and maintained with much care. The ahistorical time of progress also looks different from the end of the world. In one glance one can see a condensation of half a century of Spain's energy history, but this history does not appear as a chain of events, a glorious history culminating in a coming transition. With his back to his village, Bernat saw, like Walter Benjamin's angel of history, the piling up of rubble on top of rubble.[12]

From the place where the world ends, the abstract space and history-less time of energy looked like fetishes concealing myriad social relations. This book is written from that vantage point, one that allows at once for the enabling of entanglements and the critique of political economy.

The Old is Dying and the New Cannot Be Born[13]

Processes of energy transition, rather than energy in the abstract, are central to my inquiry. In recent decades, peak oil and climate change have made energy increasingly visible, signaling the beginning of the end of modernity's spell of energy invisibility. Peak oil and climate change cannot be brushed away as mere "malfunctions"; rather, they index the likely terminal crisis of an energy system built on fossil fuels, suggesting that high-energy modernity is but a transitory phase in human history—a "fossil interlude," in Alf Hornborg's expression.[14] We are in the midst of an energy transition. Although the outcome of this transition is uncertain, we can affirm that it will likely be influenced by how we are able to imagine it.[15]

If the fossil fuel era is dying, old habits of thought remain strong, as evidenced in the way dominant imaginaries circulate around a transition toward renewable energy.[16] Energy transitions tend to be understood in narrow terms, as mere shifts in the sources and technologies of energy supply. From this mechanistic perspective, renewables point toward a prefigured future in which some envision a more just society where centralized power, accumulation for the sake of accumulation, and widespread unevenness become things of the past. Others believe it spells the end of civilization and democracy as we know it. And for many others, still, it signals the dawn of a new brand of smart capitalism, a development that will be sustainable (respectful of the environment) and sustained (ever reproduced and perfected).[17]

These different visions share the belief that energy transitions are largely determined by energy sources and technologies. That is, the competing ideas all participate in the fetishism of the ideology of energy, a narrow, abstract understanding of energy that alienates it from the cultural mediations and societal arrangements with which it is entangled. As Abramsky and De Angelis write: "While some kind of transition to postpetrol energy sources is virtually inevitable, the form it will take is far from a technical inevitability. Rather, any transition will be the result of an uncertain and lengthy process of collective struggle, as will its qualitative aspects."[18]

This book challenges the idea that renewable energy necessarily involves a stark rupture with former modes of energy production. The adoption of renewables may give rise to different forms of energy transition involving varying degrees and forms of change and continuity with the existing energy system. My ethnography shows that energy transitions are multilayered processes that open possibilities for new social arrangements, while also highlighting the ways that such new social arrangements rework inherited power dynamics.

The central part of this book (chapters 4–7) is devoted to the development of wind energy in Spain. Today, wind is the main source of electricity in the country. Analyzing the institutional arrangements, cultural mediations, power structures, and social relations of production through which the energy from wind is harnessed, I unveil the contradictory tensions that pervade this process, which are intimately linked with the continuity of preexisting patterns of organizing energy supply.

These continuities are evident in the organization of production, based on the reproduction of a centralized structure controlled by a handful of Spanish transnational corporations, and in the relations of production, marked by an extractivist logic and the peripheralization of producing regions such as Southern Catalonia. In Spain, the development of wind energy has been largely dominated by the oligopolistic utilities that have traditionally controlled the electric sector, largely setting its timing and direction, including the paralysis that has affected the sector since 2013. Rather than contributing to the creation of a more sustainable economic base, wind energy was placed in service to a structure of accumulation that was intensive in the use of materials and energy. Community involvement in wind energy development has been almost completely absent, triggering the alienation of local populations.

The promise to create a more distributed energy system has largely been squandered, with wind farms clustering in impoverished rural regions, replicating the unequal ecological and economic exchanges that have characterized

Spain's electric system since the early twentieth century. These nefarious consequences have largely deactivated the transformative potential, and limited the environmental benefits, of wind development. But it wasn't inevitable—one need only look to Denmark and Germany to see other European wind energy leaders purposefully taking up a different path (see chapter 5).[19]

Regional Ethnography

Analyzing energy transitions anthropologically involves situating these transitions in time and space. As Dracklé and Krauss have argued, the transition toward renewable energy is a global process that occurs locally: "The global energy transition occurs in specific settings with their own histories of people, common laws, rules of ownership, and material cultures. This process encompasses a transition of power as energy and as political power relations, thereby making possible an anthropology of energy."[20] The localized character of specific processes of energy transition makes it not only possible but also necessary to study them ethnographically.

The setting of my ethnography is Southern Catalonia. The region has a total population of around 200,000, and is composed of the five southernmost counties (Terra Alta, Ribera, Priorat, Baix Ebre, and Montsià) of Catalonia, located in the northeastern corner of Spain (see map 1).[21] Although Catalonia is one of the wealthiest *comunidades autónomas*—the first-level administrative unit in Spain—and is usually considered the industrial engine of the country, Southern Catalonia offers a different perspective. It is the poorest and least industrialized region in Catalonia, with family income at 75 percent of the Catalan average,.[22]

Southern Catalonia is, thus, a periphery, a lagging rural territory roughly equidistant (around 200 kilometers) from three of the main Spanish metropolitan regions: Barcelona, Valencia, and Zaragoza. Most of its population is rural; the only city, Tortosa, has less than 30,000 inhabitants. Although Tortosa is widely considered the capital of the region, it does not perform the function of an urban center for many Southern Catalans, who prefer to go outside Southern Catalonia—to Tarragona (the provincial political capital) or Reus (the provincial commercial capital)—for services (lawyers, doctors, shopping, etc.).

Despite the region's weak internal integration, Southern Catalans have a remarkably strong sense of shared identity. Folklore and dialectal traits are identity markers distinguishing Southern Catalonia.[23] And the region's shared history is profoundly marked by two events: the Spanish Civil War (1936–1939), along with the ensuing repression, which included especially bloody episodes; and the rural exodus of the 1960s and 1970s, the paradigmatic symbol of the collapse of an old agrarian order.

My regional focus allows me to situate the development of renewable energy within a broader historical framework. Indeed, during the 1970s and 1980s,

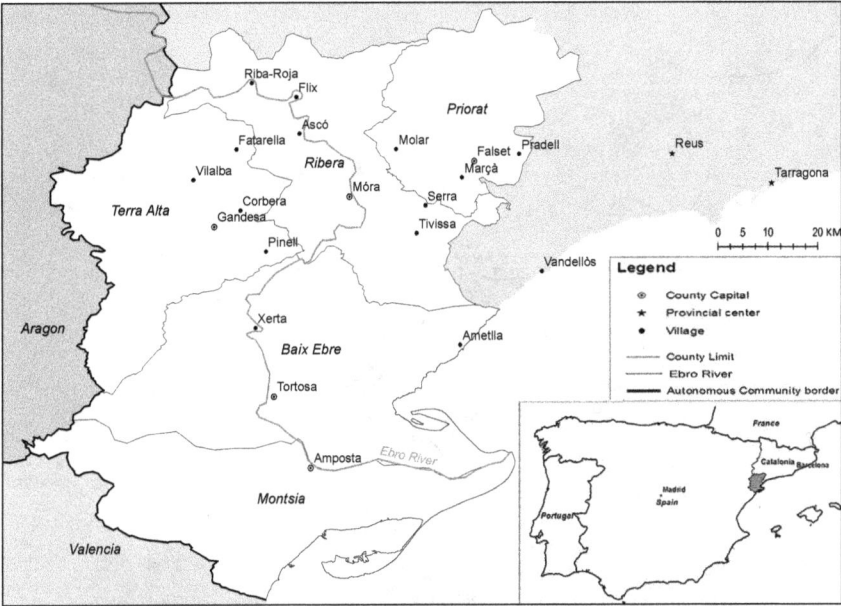

Map 1. Southern Catalonia.

Southern Catalonia was at the center of an attempt to develop an energy transition toward nuclear energy (map 2). Four of Spain's ten nuclear reactors were located here. In comparison, the role of the region in the development of wind energy has been modest. The sixteen wind farms installed in the region, totaling around 500 MWh of installed capacity, represent a small portion of the nearly 23,000 MWh installed in the whole country.[24] Nonetheless, Southern Catalonia's role in wind energy development should not be underemphasized: it accounts for half of the installed capacity in Catalonia despite representing only 10 percent of its surface and barely 3 percent of its population.

The coexistence of two energy transitions, to nuclear and to wind, provides a privileged opportunity for examining the changes and continuities that characterize the history of the Spanish energy system. And, as one could appreciate from the place "where the world ends," the region hosts myriad energy facilities, from high tension lines and pipelines to solar farms, natural gas power plants and hydroelectric dams (see map 2). What one *sees* from "the end of the world" is just a fraction of the projects that have been slated for the region. It is only stern local opposition that has prevented the construction of more energy infrastructure, including two dams, three nuclear plants, scores of wind farms, and two large storage facilities, one for natural gas and the other for nuclear waste.

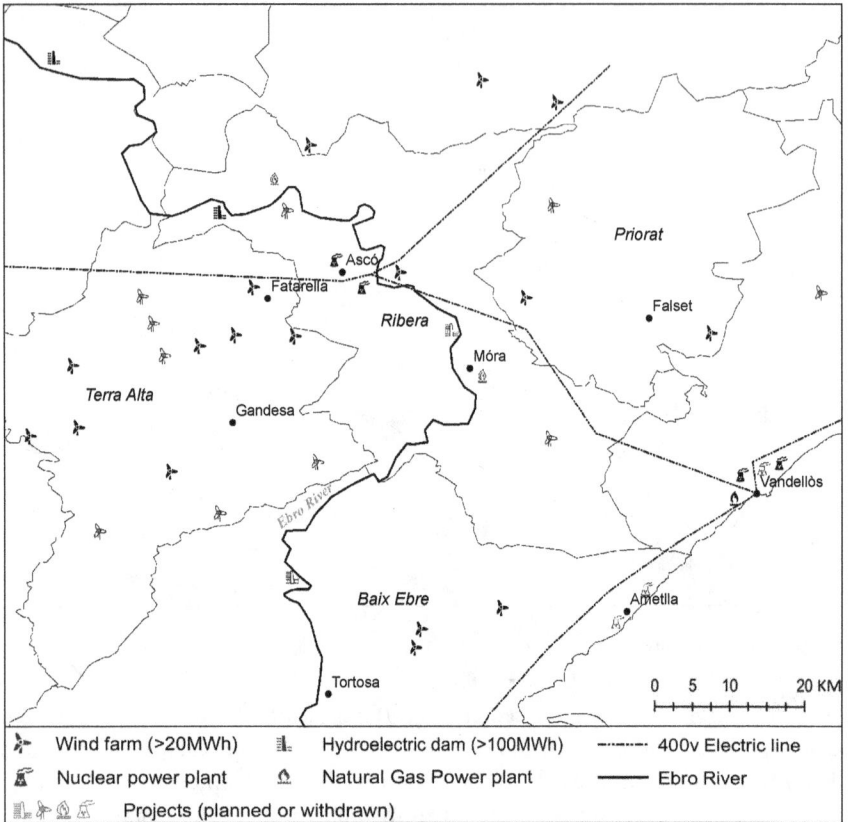

Map 2. Main energy infrastructure in Southern Catalonia.

Most of these infrastructures are located (or were to be located) in the northern half of the region, made up of the counties of Terra Alta, Ribera, and Priorat. With a total population slightly under 50,000, these three counties are the least populated and most agrarian of Southern Catalonia. They specialize in dryland farming, and only Ribera possesses a modest industrial base. Although widely practiced, agriculture is rarely the sole source of livelihood for the inhabitants of these counties, whose domestic economy typically relies on a variety of income sources. Commercial and kinship ties between what the locals often call *les tres comarques* (the three counties) are strong. They have lived with energy production longer and more intensively than any other part of the region, and they have the most marked trajectory of opposition to energy facilities.

I conducted eleven months of discontinuous fieldwork between 2010 and 2014. During this time, I lived in the village of Fatarella (population 1,000). Much of the ethnographic material presented in this book was gathered there. Fatarella is strategically positioned at the center of the Southern Catalan electric hub: less than ten kilometers from the hydroelectric dam of Riba-roja and the two nuclear reactors of the village of Ascó, it is currently surrounded by three wind farms. It is through living in Fatarella—accompanying the locals to their fields, participating in their *fiestas*, listening to their stories about the arrival of the nuclear plants and their work in them, trying to understand why they thought that the wind farms had been a "bad deal," paying attention to the aspirations of women and men and the ways they made their livelihoods—that I learned how Southern Catalans engage with energy, both as a concrete abstraction and an all too quotidian reality.

My fieldwork was not limited to Fatarella, however. I spent a good deal of time on the small roads of the three counties. In one of the paradoxes of fieldwork, my research on energy committed to a transition away from fossil fuels involved burning quite a bit of gas. Only thus could I reach the villages of Southern Catalonia and conduct a regional ethnography. I interviewed activists and local mayors, landowners affected by wind farms, former and current workers of the nuclear plants, local intellectuals, peasant leaders, and local civil servants. I also attended demonstrations, political rallies, and cultural events, participated in activist meetings, trekked through the mountains, worked in archives, and visited local wine cellars and cooperatives, major actors in the life of the region.

A third circle of ethnographic inquiry took me outside the region, mostly to Barcelona. Although I interviewed officials of the Catalan government, academics, and environmental activists, the main purpose of these trips was to get to know those involved in the development of wind energy. It took me a while to access them, confirming the widespread belief that the electricity sector is surrounded by a fair amount of secrecy. Yet the snowball rolled, and I was able to interview independent wind developers, managers and workers of electric utilities and wind turbine manufacturers, and representatives of the industry board. I wanted to understand, from their perspective, the lengthy process of obtaining the permits and the land access rights necessary to build a wind farm, the way they understood their contribution to the territories where they installed their wind farms, their relationship with politicians, and their reaction to the regulatory changes affecting the sector. Fundamentally, I wanted to learn about their work from the point of view that I developed through my local fieldwork. Things look different when you see them from a periphery.

Periphery, Waste, and Dignity

Anna Tsing argues that the end of extraction turns resource peripheries into spaces of abandonment. Southern Catalans also feel abandoned, yet this feeling is not the result of divestment, but of the fact that the region is constantly renewed as an electricity-producing periphery. Each new round of energy investment in

Southern Catalonia has iterated the region as a periphery, a place from where resources, electricity and profits are extracted, leaving behind environmental risks and social conflicts.

Capitalism is based on ecologically and economically unequal exchanges between central and peripheral regions.[25] Centers operate through the constant accumulation of order and net energy, displacing entropy to the peripheries, and thereby imposing, in Raymond Williams's words, "an at once profitable and pauperizing order on them."[26] That is, peripheries are peripheral *in* capitalist processes of accumulation, but not *to* them. Therefore, regions such as Southern Catalonia offer an advantageous perspective for the analysis of political economic structures and broader patterns of accumulation.

The Spanish electricity system materializes a social and functional arrangement of cores and peripheries. Peripheralization both enables and is, to a large extent, the result of the rent-extracting behavior of the electric sector in the larger context of the Spanish economy, allowing for the reproduction and restructuring of certain patterns of accumulation, class structures, and social relations of production. The role of the electricity system in organizing unequal relations of exchange and processes of uneven development reveals that it works as what Eric Wolf called a field of "structural power," specifying not only the distribution and direction of energy flows, but also structuring the value relations that organize the social division of labor and nature.[27]

Vinay Gidwani's theorization on the concept of *waste* illuminates these value relations.[28] Waste, Gidwani argues, needs to be understood in dialectical relationship with capitalist value: it is "the economic and moral antithesis of value," identifying all that which is stagnant, retrograde, pointless, worthless and, therefore, a threat to the order established by capitalist value relations.[29] In Southern Catalonia, every new energy project has been justified by presenting the region, its inhabitants, and their practices as waste: recessive agriculture, lagging development, absence of profit-making activities, and so forth. Each new round of energy investment has thus been presented by the state and the energy sector as an opportunity to redeem the area from its underdevelopment. Yet with every round, the region has continued its trajectory of depopulation, impoverishment, and agricultural decline. Energy development adjusts the region to the law of value, yet in that process, any practice that does not serve the energy sector—local attachment to land, the efforts to produce autonomous, self-managed livelihoods—is iterated as waste.

And yet, waste actually *is* productive of value. Places like Fatarella were attractive to wind developers not only because of the untapped potential that traverses the area (wind), but also because the land has been conceived of as disposable. Southern Catalonia's valuelessness is a function of the dialectic between centers and peripheries structuring the electric system as a field of power, and it has allowed developers to make big profits by paying little for that land. That is, devaluation works as a precondition of capitalist expansion: because certain places and peoples are constructed as waste—residual, barren, marginal, disordered—capital can justify the need to intervene and make them valuable, that is to say, value-producing.

Since the 1970s, Southern Catalans have opposed their peripheralization and every new energy project with a demand for dignity.[30] Dignity should be understood as the central element of a local "theoretical framework"—to use Susana Narotzky's expression—aiming to explain but also to disrupt the value relations that both sustain and result from a particular political economic structure that allows for the extraction of profits from the area, producing its inhabitants and their possessions, most notably land, into waste. With every new round of energy investment, Southern Catalans have two basic options: to surrender to the logic of the state and energy corporations, accepting that they are waste, or to fight against it, demanding their dignity.

This demand for dignity takes two different, although often overlapping, forms. First, it emerges as indignation, a fiery reaction against passively accepting the denial of one's own dignity. In this respect, dignity is the contrary of resignation and deference. It is also a refusal "to conceive of oneself as someone else": accepting the idea that one is waste amounts to accepting the idea that one wants to become someone else.[31] The second form emerges as an assertion of dignity, understood here as worth. It claims the value of Southern Catalans and their possessions, especially of their land, and of the region as a whole. Crucially, it asserts the value of the relationship between people and place over time, a link that is at the basis of the construction of the region as waste, but also of the reproduction of local livelihoods. Thus, as I argue in extenso in chapter 7, dignity emerges not only as the opposite of waste, but as the *other* other of value.

Dignity, write Bonefeld and Psychopedis, is "rightly seen to resist the full utilization of technical efficacy of social labor power and its transformation into an effective and compliant resource that feeds the well-oiled systems of economic production and political domination."[32] It demands social relations in which humans recognize themselves as having a purpose—not as mere resources or means, but as the subjects of their own social world. The demand for dignity appears, in this respect, as what Jane Collins calls a "revaluation project," seeking to adjust or reconfigure value relations to the long-term reproduction of local livelihoods.[33] It expresses a set of alternative values to a capitalist law of value that condemns the region and its inhabitants to chronic peripherality. These values emerge out of the daily struggle to make a living in and from the land, in the region.

The demand for dignity indexes the effort to reproduce a struggle that occurs daily, a search for autonomous livelihoods rooted in the peasant struggles of the early twentieth century. It is through this daily struggle that Southern Catalans build a precarious outside to the value structures of Spanish capitalism and the power of the energy system. In this way, the demand for dignity asserts the incommensurability between the two worlds that Bernat showed me from the "end of the world."

With their claim to dignity, Southern Catalans affirm the value of their livelihoods and resist a historical process of marginalization. But they also contest the increasing centralization of Spain's energy model and the incumbent concentration of political and economic power. Dignity not only encapsulates the cultural codes through which economic and ecological imbalances are locally understood and resisted. It also contains social and political possibilities for counter-hegemonic understandings of energy transitions and the current political economic conjuncture in Spain and Europe.

Notes

1. On the invisibility of energy, see Boyer (2011), Shever (2012), Huber (2014) and Hughes (2017).
2. Love and Isenhour (2016); see Cottrell (1955) and Nye (1999) for similar formulations.
3. Smil (2008: 380).
4. Nader (2010: 540).
5. Mitchell (2011: 231).
6. On this connection, see Adams (1975), Debeir et al. (1991), Boyer (2014), and Malm (2016).
7. Horkheimer and Adorno (2002: 2; my emphasis).
8. On concealment as a central attribute of power, see Bourdieu (1977); see Martínez Alier (1987) on the contradiction between economic growth and biophysical processes.
9. Tsing (2015: 5–6; my emphasis).
10. Kalb and Tak (2005). On the entanglement of *naturecultures*, see Haraway and Wolfe (2016). For recent programmatic approaches to the anthropology of energy, see Reyna and Behrends (2011), Strauss et al. (2013), Boyer (2014), and Love and Isenhour (2016). On fetishism, see Friedman (1974), Taussig (1980), and Hornborg (2016).
11. On the abstract space of capitalist extraction and accumulation, see Lefebvre (1991), Cronon (1991: 23–97), Watts (2004), Ferguson (2005), and Malm (2016: 298–307). On the temporality of capital and the simultaneity of the nonsynchronous, see Bloch (1991), Lowy (2005), Anderson (2010), and Harootunian (2015). On the temporality of the quotidian in the Mediterranean, see Bourdieu (1977), De Certeau (1984), Herzfeld (1991), and Lefebvre (2004). On household names, see Zonabend (1980).
12. Benjamin (1968a, Thesis IX).
13. Gramsci (1971: 276).
14. Hornborg (2013); see also Pomeranz (2000) and Sieferle (2001).
15. Urry (2014).

16. See Podobnik (2006), Abramsky (2010), Nader (2010), and Smil (2010) for insightful takes on energy transitions.

17. Scheer (2006) and Mitchell (2011) constitute examples of the first two views; the third one is associated with "ecological modernization," discussed in chapter 5.

18. Abramsky and De Angelis (2009: 9).

19. For an insightful comparison of European cases, see Szarka (2007).

20. Dracklé and Krauss (2011: 5); see also Bridge et al. (2013).

21. The geographic boundaries of Southern Catalonia are not clearly established. These boundaries were the object of intense debate, especially during the 1970s, the period in which the historical demand for regional self-government reemerged. The main point of disagreement had to do with the inclusion of Priorat (Bayerri 1985 and Alonso 1978: 185–190). In 2001, the Catalan government created a new administrative region (*Terres de l'Ebre*) that did not include Priorat, thus comprising only four counties. However, given the strong connections that Priorat maintains with Ribera and Terra Alta, I have opted for including Priorat in my definition of Southern Catalonia. I have also included the southern villages of Baix Camp (primarily Vandellòs and L'Hospitalet de l'Infant) as part of Southern Catalonia because, administrative boundaries notwithstanding, their inhabitants identify themselves as belonging to Baix Ebre.

22. Data on Brute Family Revenue, last modified April 10, 2017, accessed October 3, 2017, https://www.idescat.cat/.

23. For an overview, see Guiu (2008).

24. AEE (2016).

25. See Gill and Kasmir (2016) and Werner (2016) for recent discussions on uneven development.

26. Williams (1980: 79).

27. Wolf (2001).

28. Gidwani (1992, 2008, 2012). In addition to Gidwani, my take on value draws primarily on Elson (1979), Cleaver (1992), Graeber (2001), De Angelis (2007), Burawoy (2013), and Collins (2017); for an ecological perspective on the subject, see Altvater (1993), Coronil (1997), O'Connor (1998), Foster (2000), Hornborg (2011), and Moore (2015).

29. Gidwani (2012: 277).

30. For recent explorations on the idiom of dignity in Spain, see Narotzky (2016a and 2016b) and Franquesa (2016).

31. Bromell (2013: 289).

32. Bonefeld and Psychopedis (2005: 4).

33. Collins (2017).

1 Dependence and Autonomy

The peasant utopia is the free village, untrammeled by tax collectors, labor recruiters, large landowners, officials.

　　—Eric Wolf, *Peasant Wars of the Twentieth Century*

The peasant-utopia has ... never existed. There has never been a peasant-community completely free of taxes, rents, fees, tithes or labor-services, a "free village" in which smallholders have had absolute security of tenure and freedom from subjection.... The aspirations of peasants for this kind of independence have, however, made themselves felt in various ways—in peasant-rebellions, in political and religious movements, and in cultural traditions.

　　—Ellen Meiksins Wood, *Peasant-Citizen and Slave*

DURING MY FIRST visit to Southern Catalonia, on a sunny spring morning in 2010, I was struck by the impression of a farmland without farmers. The stone walls and terraces filled with vineyards, olive groves, and hazelnuts bore unmistakable signs of human action and appeared to be in full production, yet this landscape was remarkably empty of people. Indeed, the most visible active presence was that of the groups of workers involved in two infrastructural projects: the construction of wind farms and a government-funded, countywide irrigation project.

Within a few weeks, I discovered that in the evenings and on weekends, those empty fields acquired new life: men and often entire families tending their fields and gardens with small tractors, or socializing, mostly during the weekend, in their *masos*.

"In the city they go to the gym after working, here our workout is going to the fields," people often told me. During those first weeks, in my conversations with local residents about the wind farms, I often interjected questions about the irrigation project, conceived in the official jargon as a "system of support"—not to promote crop replacement but to alleviate the effects of extremely dry summers on hazelnuts, almonds, grapes, and olives, the traditional commodity crops of the region. The answer was invariable: "I don't think I'll pay for the connection; it comes too late, anyway, twenty or thirty years too late. Then, it could have made a difference, now there are no peasants anymore."

This sense of despair infused Southern Catalans' views of agricultural activities, a feeling ossified in a common series of complaints: "It is impossible to make a living from agriculture," "What can you do with these prices?", "This is not the right way to farm, it is just suffering," "We will all end up living cramped into cities," "Sometimes I think they should just finish us off."

This last statement always intrigued me, not so much for the elliptical character of the subject "they"—sometimes replaced with "those in charge" (*los que manen*)—as for the ambivalence of the object "us," which seemed to invoke a nebulous collective identity, simultaneously the peasantry, the rural world, and Southern Catalans. Yet in all its fuzziness and simplicity, the statement captured not only a local experience of rural exodus and anxiety about the future, but also an awareness, however abstract, of the structural dynamics fueling that experience, which Marx condensed in a famous sentence: "the bourgeoisie ... has made the country dependent on the towns."[1]

As months went by, I also observed how this overwhelming sense of despair coexisted with more subdued yet no less unconcealed feelings of pride and joy in farming activities and the condition of *pagès*—a cousin term to the French *paysan* that translates as "peasant" or, in some circumstances, "farmer."

"Here, we're all *pagesos*," local men and women of all ages would say, suggesting the existence of reproductive logics that escape official statistics. This became clear on my visit to the field in the early fall of 2012, around harvest time, when, in addition to the usual gifts of hazelnuts, wild mushrooms, and grapes that my family and I used to receive from neighbors and friends, we were inundated by a daily bounty of garden produce—tomatoes, melons, zucchini, watermelons, eggplants, peppers of all sizes and colors—that the villagers foisted on us as we passed in the street. Carrying baskets and boxes of vegetables, they excitedly compared the treasures nourished by newly arrived water, despite the skepticism that I had previously recorded. "I don't know about the hazelnuts, because with these prices ... but in the winter, the pantry will be a joy!"

These feelings and responses, contradictory and oblique, are the structurally contingent product of a long history of material relations—institutionalized in local culture, imprinted in historical consciousness, and inscribed in the landscape—that permeate daily practice. At the center of this historical process we find the "crisis of traditional agriculture,"[2] a term that identifies the obverse of the modernization process initiated by the Spanish economy in the 1960s, often known as the "Spanish Miracle." Depeasantization, massive rural exodus, and a structural crisis in family farming became the negative image of industrialization, urbanization, and extension of wage relations. Rural areas and small farming were sacrificed on the altar of growth and economic development.

This chapter situates this epochal transformation within a broader historical analysis of the development of material relations in Southern Catalonia since the late nineteenth century. Adopting such a long temporal perspective makes it possible to examine historical dynamics governing the transformation and continuity of the Southern Catalan rural world. This examination will focus on the *articulation between two problems*: on the one hand, reproduction, or the drive of people to reproduce their livelihoods and how they understand their actions; on the other hand, production, or the drive of capitalists for accumulation and how they try to manage other people's lives to that end.[3] The history of the Southern Catalan countryside may be described as a permanent struggle to reproduce autonomous livelihoods in the face of manifold political and economic structures producing dependence and marginalization.[4] On some occasions, this struggle emerges as open conflict. More often it manifests in daily practices of carving out autonomy, in the stubborn unwillingness to be finished off, and in identity claims that escape capital's law of value. Only through this historical lens, attentive to the precarious, shifting balance between autonomy and dependency generated in the dialectic of production and reproduction, can we understand the rejection of energy facilities in Southern Catalonia.

Dependence, Struggle, and Violence in Southern Catalonia (1874–1939)

Southern Catalonia is an area of dryland farming of Mediterranean commodity crops, predominantly grapes, olives, hazelnuts, and almonds. Specialized agriculture has been a characteristic feature since the eighteenth century, when the region tripled its population and agrarian production soared.[5] Throughout the nineteenth century, as Catalonia became the first Spanish region to industrialize (especially along the coast), the livelihoods of the Southern Catalan peasantry progressively deteriorated, principally as a result of the worsening of the terms of trade, rising taxation, and the state-led privatization of common lands.[6] These factors combined with increased demographic pressure to produce heightened class differences. My description of the system of dependencies structuring the Southern Catalan "traditional agrarian society" focuses on this period, which politically corresponds with a regime known as *Restauración* (1874–1931). During this period, successive governments dismantled the democratic reforms introduced between 1868 and 1874 and implemented an infamously fraudulent electoral process that effectively reinstated the power of the landowning elites at the national and local levels.

Nested Dependency Ties

The settlement pattern in Southern Catalonia was—and remains to this day—quite uniform, creating a structure that may be described as an archipelago of villages. Southern Catalans lived in small- and medium-sized villages, generally between five hundred and two thousand inhabitants, separated by regular distances (five to ten kilometers). Villagers rarely owned land outside of their *terme municipal* ("municipal territory," the area administratively controlled by the municipality), creating a strong correspondence between landownership borders and administrative units. Dispersed habitation was uncommon, and distance to the agricultural fields—with each household typically cultivating several plots—was often considerable, a circumstance that helps explain why "farming" is locally referred to with the expression *anar al defora* ("going outside"). With the exception of the Ebro riverbanks, the terrain is rugged, dry, and full of stones, and has historically been made arable through the construction of a complex system of terraces and stone walls. The area was isolated, with the main communication lines being the river Ebro—linking the area with Tortosa, the region's historical capital and its only sizable town—and a railroad to Reus—the trading market for the region's agrarian production—built in the 1850s.

Each village in Southern Catalonia may be described as a constellation of *cases* (literally "houses," yet roughly translatable as "households" and "farmsteads") operating as units of both production and consumption.[7] The region thus offered a fairly typical picture of petty commodity producers combining the sale of agricultural commodities with a noncapitalist organization of the labor process. Every household complemented the production of commodity crops with self-provision—raising small animals and pigs, growing fodder for mules, tending vegetable gardens and fruit trees at the edges of the fields—and exploitation of forested areas for wood, herbs, mushrooms, and, probably most important of all, hunting.[8] The *casa* operated as a central cultural category evoking cooperation, affection, and unity of purpose, playing a central role in the definition of a person's place in society. Thus, to this day Southern Catalans tend to hypostatize the *casa* as a subject of social relations, saying, for instance, "*casa* X married *casa* Y," or "Joana is a sister of *casa* Z."

The notion of *casa* thus possessed a strong ideological component stressing its organic unity, thereby obscuring the fact that the relation to the means of production within the household (fundamentally, landed property) was not collective or homogeneous. Indeed, the ideal of undivided inheritance through primogeniture structured a double axis of inter- and intragenerational contradictions. The former tended to be expressed in conflicts over decision making within the *casa* between the parents of the heir (*amos*), who tended to retain ownership of the estate until their death, and the heir and his wife (or the heiress and her husband). Within the same generation, lines of power and economic prevalence set cleavages between the prospective heir and his or her siblings, who could only remain in the *casa* as unmarried dependent laborers. Marriage decisions were crucial for the reproduction of the *casa*, and they tended to be monopolized by the *amos* and the married heir.[9]

Moving from the social relationships within the *casa* to those between *cases*, we observe that differential access to landed property was key to the hierarchy of *cases* and the relations of dependence between them. At the top of the pyramid we typically encounter a small number of relatively large landowners (*cacique* or *senyor*) whose land was not cultivated directly, but by poor peasants in share-cropping arrangements. *Senyors* not only possessed more, but also better land, situated on the plains closer to the village. They also enjoyed a position of semi-monopolistic control over the transformation and commercialization of the agrarian produce. Their control over the life of the community was reinforced by their tight grip over the electoral processes and through the support of the church, the administrative apparatus, and the repressive forces. These landowners were the nodal points in a complex structure of power, known as *caciquismo*, that blossomed during the *Restauración*. Joaquín Costa described it in 1901:

> Each region ... was dominated by a *cacique*.... To know how a legal matter would be resolved, it did not matter whether you were right or whether the law was on your side: you had to ask whether the *cacique* ... was on your side or not. He determined who had to go to the army ... which letters got lost ... who was fined ... how much taxes every person had to pay ... [and whether] roads were built to go to his estates. He had a whole territory ... divided into smaller territories controlled by lesser *caciques* ... forming ... a thick fabric that encompassed the whole country.[10]

Some years later, the politician Manuel Azaña stressed the personal structures of dependence underpinning the broader networks described by Costa: "It is a mistake to see *caciquismo* as a recent invention of our electoral industry.... Without elections there would be the same oppression.... [A]t the feet of the cacique there is always a group of people without freedom. They will not be redeemed with a simple electoral law."[11]

Indeed, "at the feet of the *cacique*" we find a large number of land-poor and landless *cases*. Poor *cases* could only manage their chronic uncertainty by resorting to a combination of dependency and labor mobility.[12] The main form of dependence was to become a *mitger* (sharecropper), a condition that immersed the poor peasant in a closed world of loyalties and personal relations dominated by the landowner.[13] Thus, among other things, the peasant would have to show the cacique respect and call him *senyor*, vote for the party the *senyor* supported, send his children to the farmstead school, and participate in the festivities of the *senyor*'s *casa*. Poor *cases* combined sharecropping with other labor strategies, notably sending some of their members to work as servants (*mossos*) in wealthier households, often outside their village, or to work for a wage (*jornalers*) in peak periods of the agricultural cycle—for instance, during the olive harvest, a task predominantly performed by women. Poor peasants tended to engage in wage labor only during certain phases of the life cycle, typically before marriage.

The main class antagonism in the Southern Catalan countryside was, therefore, not between landed farmers and landless workers, but between large landowners and a precarized mass of peasants—the members of poor *cases*—that combined sharecropping and the sale of their labor with the cultivation of their own insufficient land.[14] In between the large landowners and the small *cases* we find a third group of middle *cases* (*cases mitjanes*). Although their situation was very variable—with some at permanent risk of slipping into a position of dependence and others in a powerful position (*cases fortes* or "strong *cases*")—middle *cases* were characterized by possessing enough land and labor (through kin and non-kin dependents or *mossos*) to make an independent living. They most closely embodied the ideal of the *casa* as an autonomous unit of production and consumption.

Three corollaries emerge from this description. First, the Southern Catalan social structure as it existed at the turn of the twentieth century was articulated through a series of personal dependency ties—between landowner and sharecropper, small producer and trader, inheriting and non-inheriting siblings, *cacique* and everyone else. These ties, using Godelier's expression, "functioned as social relations of production": they organized the labor process, regulated access to the means of livelihood, and organized the distribution of wealth and power.[15] Economic exploitation and personal subordination were inextricably intertwined. Second, social relations of production were wrapped up in ideological discourses stressing unity and pervaded by affective qualities. This was reinforced by the strong interweaving of intra- and inter-household relations of production, as Terradas stresses: "Agricultural servants lived in peasant houses that had the same interests as that of their own parents and with a way of working where obedience could not be separated from contract nor loyalty from responsibility, nor—and this might be the most important—ingratitude from revolt."[16] Third, although the affective qualities attached to relations of dependence and exploitation tended to obscure power divisions and class contradictions, they did not preclude conflict. Peasant struggles against economic exploitation necessarily contained a strong element of rejection of personal forms of authority, servility, and deference and, by extension, of the discourses and institutions that sustained their legitimacy.

Conflict, Violence, War

During the last decades of the nineteenth century, peasant resistance in Southern Catalonia took on a variety of localized or spontaneous forms—such as defense and occupation of communal lands and violent, mob-like actions against village officials and state representatives. In the first third of the twentieth century, these forms of resistance were progressively superseded by three principal modalities of stable, organized peasant action.[17] First were actions demanding better conditions for agrarian wage labor, such as the eight-hour workday. Socialist and anarchist unions led these forms of action, which frequently included strikes. Second was the creation of local cooperatives for the transformation and commercialization of wine and olive oil. Small and middle peasants were the main promoters and beneficiaries of these cooperatives, which allowed them to circumvent the control that landlords and traders, historically the owners of processing mills, had over the commercialization of agricultural commodities.[18] Finally, a demand for access to land ownership, synthesized in the claim "land for the tiller" (*La terra per qui la treballa*). The moderate left union Unió de Rabassaires (UR) led this struggle with massive support from sharecroppers—especially in the wine-producing counties of the Catalan South—and became the hegemonic peasant organization of 1920s and 1930s Catalonia.

The hegemonic character that this third form of struggle achieved deserves some consideration. As Izquierdo Martín and Sánchez León argue, the *rabassaire* movement was inextricably linked to the memory of the first Spanish republican period (1873–1874), during which sharecroppers obtained land ownership *and* citizenship rights such as voting: "The Catalan peasant struggle appears as a form of cooperation oriented to reproduce in the twentieth century the political conditions that half a century earlier had made possible the access of peasants to landownership."[19] In other words, the demand "land for the tiller" pointed to a particular collective memory—an "uncompleted past," in Ernst Bloch's expression—and was oriented to a transformation of the relations of production, abolishing the material conditions that kept peasants dependent and impeded their participation in political, public life.[20] The struggle for land was a struggle for political freedom and against the vertical ties that defined *caciquismo*.

The idea of freedom that emerges from this struggle is thus quite specific. It is not the liberal, abstract notion of freedom—*freedom of* religion, *freedom of* enterprise, and so on—but rather a republican notion, freedom as autonomy, that is to say, *freedom from* those relations that kept the peasantry subordinated or dependent. This republican freedom, crucially linking rights and livelihood, constituted the basis of full citizenship. It emerged as the opposite of Marx's infamous "idiocy of rural life": as the capacity to intervene in public affairs. For UR, this freedom would naturally allow the peasantry, as citizens, to establish horizontal alliances with the urban working classes, simultaneously redefining the relationships between country and city and ultimately superseding this division, thus making evident the intimate connection between the peasant and the territorial question in Catalonia. As republican geographer Pau Vila argued in 1930, the goal was not the return to a romanticized rural arcadia or to resettle the countryside, but rather to improve the living conditions of those who lived there and to *citizenize* it: "What is needed is to keep in the countryside those that remain in it. For this, it is necessary to create communications knitting villages and towns together, and that these communications bring modern comfort to the countryside. All Catalonia should be city."[21] Obtaining full citizenship rights, removing dependency ties, and abolishing the division between country and city are, therefore, intimately enmeshed in the republican demands of the Southern Catalan peasantry, building a chain of equivalences that has challenged the logic of accumulation and centralization to this day.[22]

In 1931, the electoral victory of the left and the proclamation of the Second Spanish Republic (1931–1936) provided a favorable context for the elevation of exploited peasants' interests and the implementation of agrarian reform. On the one hand, the end of electoral fraud, previously widespread in the countryside, led to the constitution of left municipal governments in most villages, weakening clientelist networks. On the other hand, the Spanish Parliament passed a series of laws improving agrarian labor conditions, while the Catalan government, led by a nonrevolutionary left political party (ERC, Esquerra Republicana de Catalunya) with strong connections to UR, passed a law (Llei de Contractes de Conreu) granting sharecroppers gradual access to landownership in 1934. The agrarian oligarchy was terrified by this evolving state of affairs and became a major social force behind the intransigent opposition of far-right political parties to agrarian reform and, later, the military uprising that marked the beginning of the Spanish Civil War (1936–1939).[23]

The military uprising was stopped in Barcelona by the CNT (Confederación Nacional del Trabajo, the main anarchist union), allowing this organization to play a dominant role in the political direction of Catalonia during the first year of the war. In that period, the region was in the rearguard of the military confrontation, and the site of numerous spontaneous acts of violence directed against "the priest, the *cacique*, the scab, and the trader."[24] These acts were part of a series of bloody episodes that would shake Southern Catalonia over a decade, leaving a heavy imprint that remains tangible to this day.

As the CNT started organizing agrarian collectivizations in most Southern Catalan villages, understood to be the linchpin of social revolution, their efforts were met with highly uneven local reactions. In some villages, the process proceeded smoothly and with a wide level of local support, yet in others it was met with resistance from sharecroppers, who hoped that the absence of their landlords, most of whom had taken refuge in regions controlled by fascist forces, would allow them to achieve ownership of the land they had continued to farm.[25]

In several villages, the tension triggered by forced collectivization scaled up into violent clashes between small farmers, affiliated with UR, and anarchist activists and sympathizing landless peasants, in what historian Josep Termes has characterized as conflicts of "misery against poverty." The most infamous of these episodes took place in the village of Fatarella in January 1937. In this village, a popular reaction against the collectivizing process unleashed a chain of events—known as *Fets de la Fatarella*—that ended with the intervention of the anarchist militia and the killing of thirty-four villagers. Thirty-five others were sent to jail, falsely accused of sparking a fascist uprising. Three months later, after the intervention of the president of the Catalan government, these prisoners were released with no charges.[26]

In 1938, with the advance of the fascist troops, the war front was set in Southern Catalonia. Over 115 days, the counties of Terra Alta and Ribera became the stage for the Battle of the Ebro, the longest, bloodiest battle of the Civil War, with upward of fifty thousand casualties.[27] During the battle, families sought refuge with relatives in neighboring counties. Afterward, many elected Republican officials and revolutionary landless peasants faced exile, prison camps or execution. In Terra Alta, which had fewer than twenty thousand inhabitants, 85 people, mostly peasants, were killed and 317 were sent to jail by the new fascist authorities between 1939 and 1941.[28] Those who were able to cheat death and exile returned to find villages shattered and fields sown with corpses, rubble, and unexploded bombs. Until the late 1950s, poor *cases* in villages such as Fatarella would gain income from the collection and sale of this scrap metal. To this day, trenches, bunkers, bones, and war rubble litter the forested areas.

Early Francoism: Autarchy and Agrarian Fascism (1939–1955)

Franco's regime is commonly divided into two phases: *autarchy*, from the end of the Civil War to the mid-1950s, and *desarrollismo* (developmentalism), which lasted until Franco's death in 1975.[29] The autarchic phase is characterized by economic isolationism and the regime's authoritarian command over economic activity, combining strict commercial protectionism and draconian control over salaries with a timid program of industrialization, mainly funded by agrarian savings captured through a series of state credit institutions. During this period, the regime mobilized a corporatist discourse romanticizing the countryside as the repository of the essences of the country and the new state: the countryside as a harmonic social environment representing "wisdom, social peace, tradition, and decency" and faced with a city representing "disorder, atheism, depravation, and Marxism."[30] Under a veneer of agrarian populism, this discourse concealed the class-divided character of the "traditional agrarian society" and the violence with which the regime suppressed peasant struggle and reasserted the power of big landowners.

Backed by strong support from the Catholic Church and the propertied classes, the new regime implemented a legal and institutional apparatus that constituted an agrarian counter-reform intended to reinstate traditional agrarian structures. The vengeful character of this counterreform is made clear in the preamble of an agrarian law passed by the regime in 1940. The law's main objective was to block sharecroppers' access to land ownership, and it maintained that social relations in the countryside had to rest on the principle of "submission," justified by the fact that "after the war, peace was automatically reestablished in the countryside, and this peace was made possible by the forced silence and submission of those who misbehaved [*se portaron mal*] or behaved with indifference, which was like misbehaving."[31]

The regime rescinded the right to strike, established draconian restrictions on labor mobility, and eliminated peasant organizations, replacing them with vertical unions controlled by local elites. Agrarian salaries decreased by 40 percent between 1936 and 1951, and labor productivity decreased with the scarcity of fertilizers and fuel for the operation of machinery.[32] An even starker wage repression in the cities led to a "reagrarianization" of the Spanish population, with the male working population in the agricultural sector rising by 700,000 during the autarchic period.[33] The regime controlled the basic means of production and the main ideological and repressive institutions, while maintaining privileged relationships with state cadres and a regulated market. Subsidized agricultural prices and an abundant, cheap labor force translated into extraordinary profits for big landlords, traders of agrarian produce, and, to a lesser extent, middle *cases*. The 1940s and 1950s were the golden age of agrarian landownership in Southern Catalonia and all over Spain. Most poor peasants reestablished dependency ties with landlords, for the only real alternatives to obedience and dependence were misery and repression.

The life of Joaquim, a poor peasant from the village of Fatarella, as described by his son Andreu (b. 1944), provides ethnographic texture to the system of dependencies that engulfed efforts to make a living in Southern Catalonia after the war. When he was just fourteen, Joaquim's parents, poor peasants with little land, sent him to the neighboring county of Priorat to work as a servant in a strong *casa*. His main job was to transport agricultural produce to the port of Tarragona for export. In the early 1930s, he went back to his village as a temporary wage laborer, working for any *casa* that would hire him. In 1937, Joaquim was one of the thirty-five villagers of Fatarella who ended up in prison for a few months as a result of the *Fets*. Once the troops arrived in the village, he took refuge with relatives in a town of Aragon. When he returned to Fatarella, his house was in pretty good condition, but neighbors had stolen all of the furniture.

After the war, Joaquim became a sharecropper: "I guess he thought it better suited to a married man, so it was then that he started working that big estate of olive trees in Riba-roja," says his son. This estate was divided into three large plots, each cultivated by a different sharecropper. Joaquim and his family lived on the estate for the whole winter and hired wage laborers, generally women, to help harvest olives in exchange for food and a salary. "During those months," says Andreu, who helped with the harvest, "we would never go to the village: no school, no church, nothing. Only the wealthy went to mass."

Joaquim and his wife Magdalena also farmed two other kinds of land. One was a small plot with olive trees, part of the communal land of the village (*comuns*), over which they had de facto usufruct rights. These commons were located in the hilliest, most inaccessible part of the municipal territory, an area that during the 1940s was episodically used by the *maquis*, the antifrancoist guerrillas.[34] The other was much closer to the village and consisted of three hectares that they owned, mostly planted with grapevines. A small part of that harvest was devoted to household consumption: "To make our own wine my father used the grapes from a small corner that was very dry and sunny and made a very strong wine. Because it had so much alcohol it always brought a good price, but my father always said that the good wine was for us, that he didn't work that hard just for the people of Barcelona to enjoy it." The rest was sold to the local trader, who was none other than the mayor. "He was a drinker and a gambler, and always had troubles with money. So one year, it must have been in 1950, he didn't pay anyone. So the year after that, my father and a friend decided to go to Ascó and sell the grapes there."

A few months later, Joaquim went to Ascó again, this time to sell wood, likely taken from the *comuns*, which in the context of autarchy and energy scarcity brought a high price. On his way back, he met what Andreu calls a "vagabond," probably a poor peasant seeking agrarian wage labor. Joaquim gave him food, and they had breakfast together on the estate of Riba-roja. Later that day, Joaquim was called into the municipality. There, he was beaten, then put inside the truck of the mayor-cum-trader and transported to Gandesa, the county capital and the headquarters of the Guardia Civil (military rural police). Joaquim was accused of aiding a *maquis*. Again, he was repeatedly beaten. The next morning, Magdalena received a phone call—which she had to take at the house of the mayor, one of the few with a telephone—from the Guardia Civil. She must pick up her husband's dead body.

As Magdalena listened, she overheard the maids of the house, giggling: "[Joaquim] complained that he hadn't been paid the grape harvest ... well, now he'll get paid, yes he will." Magdalena wanted her cousin to accompany her to Gandesa, but he was scared because he had been a member of ERC, so he sent his son. When they arrived, Joaquim was still breathing. They brought him back to Fatarella.

"It was a miracle that he survived. He lived for another twenty years, but he was never the same. He was badly affected by all that, and not only physically. My mother never again pronounced the name of the mayor, she just called him the 'criminal,' and she said it in front of everyone."

Joaquim's story shows the deep interconnection between political subjection and economic subordination. The mayor was able to use his institutional position to politically repress in an exemplary manner the person who challenged him as a trader with a monopoly over the commercialization of the grape harvest. Throughout the region, authoritarianism and bureaucratic centralization reinforced local clientelist networks dominated by big landowners and big traders. Local notables' monopoly control over the local institutional apparatus of the regime and their connections with the state and the market added to their control of the means of production and gave them the capacity to undermine poor peasants' means of livelihood. As Jesús Contreras explains, "the sharecropper who wanted to distance herself from the system of loyalties [in which she was immersed] ran the risk of 'being asphyxiated,' of finding 'all doors closed,' idioms used to express the maximum degree of repression by the dominant group."[35]

Taken together, the political and violent repression that marked the Spanish postwar period extended the servant-master logic of sharecropping to the whole of civil life in the Spanish countryside. Authority and wealth, economic exploitation and political subjugation, loyalty and obedience—they all wove together to establish a hierarchical system of access to power and resources. By the 1950s, however, this system would start to be eroded by the same political regime that had acted as its guarantor.

The Miracle: Modernization and Rural Exodus

The economic results of the regime's autarchic policies were disastrous: by 1950, the country had a lower GDP than in 1935, while, in the rest of Western Europe, GDP had risen by 30 percent.[36] Throughout the decade, the two roots of autarchy—the political ideals of the fascist regime and the country's isolation from the international community—were gradually weakened. These transformations must be situated in the context of the Cold War, which positioned fascist Spain as a strategic ally of the US and the Western bloc.[37] In 1953, Spain signed a secret trade deal with the US, and two years later the country was accepted into the UN. This gradual process of international recognition was paralleled by a progressive economic liberalization, symbolized in the ascent of a new bureaucratic elite that came to be known as the technocrats.

The final blow to autarchy came in 1959 with the passage of the Plan de Estabilización (Stabilization Plan), a program of structural adjustment reforms—incentivizing foreign investment and liberalization of trade—that inaugurated *desarrollismo*. The Stabilization Plan established the basis for the "Spanish Miracle" of the 1960s—a decade of enormous economic growth in which Spain became, in the regime's prideful boasting, the tenth largest industrial nation in the world. With a massive increase in foreign direct investment—including generous loans from international financial institutions, notably the World Bank—extensive emigration to Western European countries, and the takeoff of the tourist industry, the technocratic economic policies of the 1960s and early 1970s put in place a series of Planes de Desarrollo Económico y Social (Development Plans).[38] These plans aimed at creating a process of rapid industrialization, concentrated on certain urban centers that were targeted as poles of economic growth through the development of heavy industry, including steel, automotive, chemical, and oil refinement.[39] The small metropolitan area of Tarragona-Reus was one of these poles, hosting a petrochemical industrial complex that soon grew to become—and still is to this day—the largest of its kind in Southern Europe.

Desarrollismo thus appears as a technocracy-led project aimed at enhancing national productivity, conducted under the tutelage of Western powers, and oriented to "catching up" with Western European capitalist regimes. In line with these regimes, Spanish *desarrollismo* enrolled the population in the productivity project, a process manifested in the extension of wage relations, the birth of new urban classes, and a new consumer culture.[40]

The effects of the Spanish Miracle on the countryside were profound and multiple, yet none is as visible as, or conveys a sense of crisis comparable to, the rural exodus from the mid-1950s to the mid-1970s. In the 1950s, this process overwhelmingly affected landless peasants, especially numerous in the southern half of the peninsula. Until the early 1960s, agrarian policies—wrapped, as we have seen, in a corporatist, populist discourse—did not change much. In the north of the peninsula, protectionist measures, low (though already rising) labor costs, and subsidized agricultural prices gave rise to a period of relative stability, benefiting big landowners and consolidating the livelihoods of medium peasants. This changed with the first Plan de Desarrollo (1964–1967), which stated that agrarian policies had to promote "the transfer of workers [*población obrera*] from the countryside into other sectors," thus calling for the generalization of a rural exodus that would end up affecting all sectors of the rural population.[41]

The removal of agrarian protectionist measures, together with increasing capital needs to buy industrial inputs (machinery, fertilizers) and the dramatic worsening of the terms of trade for agricultural products, plunged small and medium farmers into a situation of structural crisis, throwing a large proportion of them into the new working class suburbs. The effect was felt most profoundly in areas of dryland farming such as Southern Catalonia.[42] The official figures are staggering: the percentage of active population working in the agrarian sector in Spain decreased from 49.6 percent to 9 percent between 1950 and 1993. In Catalonia, the absolute numbers went from three hundred seventy-five thousand in 1900 and three hundred thousand in 1950 to 67,000 in 2005, with an especially sharp decline of more than one hundred thousand from 1966 to 1976.[43]

The Country and the City

Marx argued that the separation between country and city stood at the basis of the social division of labor: "The *foundation* of every division of labor which has attained a certain level of development, and has been brought about by the exchange of commodities, is the *separation* of town and country. One might say that the whole history of society is summed up by this antithesis."[44] Taking this point of departure, a long intellectual tradition, from Gramsci to Raymond Williams, has argued that the division between country and city materializes a class-inflected history of accumulation, centralization and dependence, iterated at multiple scales.[45] The modernization of the Spanish economy—as well as its obverse, the "crisis of traditional agriculture"—may be read as a rearticulation of this division. This rearticulation can be appreciated in the changing positions of agriculture and industrial manufacturing within the country's structure of accumulation. During the autarchic period, in the 1940s and 1950s, agriculture absorbed population and was a source of capital, siphoning to industry the savings of the propertied rural classes that had been made possible by the dismal salaries paid to agricultural workers and tenants. In contrast, with *desarrollismo*, in the 1960s and 1970s, the rural world became a provider of labor power through population exodus, while agriculture, increasingly mechanized, became a consumer of manufactured goods.[46]

The rearticulation of the division between country and city may be seen as a generalized episode of primitive accumulation. The twin crises that affected the Spanish countryside—rural exodus and crisis of family farming—constitute the two sides of this process: creation of a new industrial workforce and erosion of the conditions of reproduction of the small agrarian producer.[47] Indeed, for those who did not emigrate, the "crisis of traditional agriculture" was an attack on their autonomy, leaving them in a novel position of dependency. By the late 1960s, the agrarian revenues of Spanish farmers were already lower than their consumption, an untenable situation that illustrates the reliance of farming households on nonagrarian sources of income. Even successful family farms were confronted with the fact that agrarian profits did not augment at the same pace as industrial salaries.[48]

A new set of cultural representations reinforced this process of dispossession. During *desarrollismo*, the new urban middle classes, created by the process of modernization, became the model of the productive moral citizen. The SEAT 600 symbolized this whole process: this tiny utilitarian car became the emblem of the "happy '60s," a symbol of the prosperity and newfound "emancipation" of the new urban classes. The countryside offered the inverted image. If until the 1950s the official discourse had posited the rural world as the reservoir of national virtue in opposition to the "corrupt city," by the 1960s the government adopted a reversed discourse that posited "peasant culture," once exalted, as a problem, a fetter to progress and economic growth that must be replaced by a modern, urban culture.[49] This symbolic degradation was palpable at the level of popular culture, and even in common parlance *pagès* became a derogatory term. Lack of services and public investment contributed to the image of backwardness and abandonment of the countryside, nurturing a stereotype internalized most keenly by the rural youth. The official discourse stressed that family farming, undercapitalized and with an important component of self-provision, had no place in the prosperous, modern, European Spain-to-be. The modernization of agriculture was to be achieved almost exclusively through the increased use of industrial inputs. Peasants had to become capitalist farmers or abandon their fields.

In Southern Catalonia, the erosion of the old agrarian social order did not give rise to a neat transition of peasants into proletarians and agriculture into a fully capitalist activity, but rather to widespread semi-proletarianization. It is this combination of nonviable agriculture and scarce employment alternatives, the impossibility of becoming either an autonomous farmer or a permanent wage laborer, the entrapment between an *uncompleted past* that never fully existed and an *impeded future* that will never arrive, that lies behind the cynical resignation "to be finished off" that we saw in the chapter's opening description. The phrase reflects a desire to be spared the schizoid suffering that comes with this situation. And yet, the crisis of agrarian agriculture also gave rise to new strategies to build autonomous livelihoods that should be situated within a longer trajectory of peasant struggle to escape relations of dependence.

A Most Singular Agrarian Reform

In 1964, a high bureaucrat of the Ministry of Agriculture referred to the transformation of the countryside wrought by the Spanish Miracle as "the most singular agrarian reform of all time."[50] Coming from a major representative of a regime that bloodily blocked any attempt at agrarian reform, it is hard to miss the cynicism. And yet, it does capture an essential element of what was happening in the Southern Catalan countryside: capital's capacity to erode sedimented orders and the emancipatory dimension of these solvent energies.

The erosion of the old agrarian order had deeply contradictory effects on the livelihoods of the different sectors of Southern Catalan rural society. This is most obvious in the intimate connection between rural exodus and the demise of big landownership. Barcelona and the metropolitan region of Tarragona-Reus absorbed most of the Southern Catalan emigration flows.[51] In its earlier stages, this process disproportionately affected the poorer sectors of the peasantry, yet it also had a dramatic "trickle-up" effect on big landed property. In every village of Southern Catalonia, one encounters stories about old *senyors* who lost their fortune during the late 1950s and early 1960s. These stories invariably emphasize personal circumstances such as family feuds—"the inheritors were all against each other"—and individual failures—"he spent all his money on women/gambling"—yet a single structural cause looms behind them: the loss of their sharecroppers and wage laborers. As Southern Catalan peasants often say, "These people just didn't know how to work their land. They had never done it and they weren't able to adapt to the new ways."

The demise of big landowners in Southern Catalonia is a clear example of the coeval temporality of a lowly productive agriculture and the more productive industrial economies of Spanish cities and surrounding European countries.[52] During autarchy, landowners built profits on favorable agrarian policies and the politically enforced low salaries of agricultural labor. There was no incentive to invest in increasing productivity through machinery, land improvement, and so forth. Once these workers moved to more productive economic sectors, however, salaries were pushed up and Southern Catalan *senyors* were "caught behind." They could have only responded through sudden productivity increases, replacing dependent labor with machinery. Instead, these *senyors* preferred moving to the city in search of rent opportunities, typically a combination of public offices and urban property management. Southern Catalan villages shrank as the social pyramid collapsed at both extremes, with the emigration of its poorer inhabitants and the decline of big landowning *cases*.[53]

In contrast to big landowners, small and middle peasants searched for new ways to make a living in and off the land, resorting to a combination of new agricultural investment and pluriactivity. Most middle *cases* engaged in moderate investment, buying machinery and parcels of land (often from ruined *senyors*) suited to mechanized farming. Despite this investment, middle farmers could hardly convert their farms into thriving capitalist enterprises. Indeed, the context of generalized salary increases and economic growth, stagnation of agricultural prices, and increasing capital requirements progressively weakened the position of middle *cases*.[54] By the late 1960s, this process was obvious. It forced middle *cases* to send some of their members in search of supplementary incomes, often through seasonal jobs in the booming construction and tourism sectors, but also as temporary wage laborers in neighboring regions of irrigated, value-added agriculture.

This new situation had deep repercussions for the values and structures of Southern Catalan society. I focus on three of these transformations—already palpable in the 1960s but unbearably noticeable in the present—that captured the structural crisis of family farming. First, with more and more members working outside of the household, we observe the erosion of the notion of the *casa* as a unit of consumption and production.[55] Second, land, the traditional source of wealth and prestige, started to be seen as a "weight" or a "burden," a circumstance that finds its quintessential expression in the difficulties that inheriting sons had in marrying.[56] Third, we observe a growing anxiety about the future, manifested in fear surrounding "generational continuity" (*relleu generacional*), be it at the level of the individual farm or household—the fear that the descendants will not continue with the farm, that the name of the *casa* will disappear—or at the level of the whole village, due to the emigration of the youngest, most able bodies and minds.[57]

For small peasants, the only alternative to emigration was semi-proletarianization. Indeed, the expanding Spanish economy afforded new economic opportunities to the members of small cases who did not emigrate. For these house-holds, farming their own land often became a secondary activity even for the head of the household, giving rise to what is commonly referred to as "part-time agricul-ture" or the "weekend peasant." Joaquim's son, Andreu, offers a case in point. In the late 1960s, when his father became too impaired to keep farming and his older sister had emigrated to Barcelona, Andreu rescinded the sharecropping contract in Riba-roja and started combining the farming of his own land with wage labor activities, a strategy that he has maintained to this day. An incomplete list of the jobs (often paid off the books) that he performed in the 1970s and 1980s include olive tree pruner, road builder, mason, welder in the construction of the nuclear plant, and foreman on a big estate of newly planted pear trees in an irrigated riverbank of the Ebro.

His main consideration in a job was that "it had to be nearby, because otherwise you feel lonely, and when you feel lonely you go to the bar and you end up spending the money that you made. I needed to be close to my kids." Meanwhile, he and his wife Margarida bought a tractor and some additional land and got rid of the vine-yard because it was too labor intensive. New varieties of olive trees brought better prices on the market. In addition to selling almonds and olives through the local cooperative, they made their own olive oil, grew vegetables, and raised a few dozen rabbits for self-provision. Moreover, Margarida occasionally engaged in domestic textile work, adding a complementary source of revenue to the household.

In short, Southern Catalonia's crisis of traditional agriculture gave rise to a series of contradictions. The upheaval of the old agrarian order destroyed cer-tain structures that provided a degree of autonomy and security. This imposed new forms of dependence even as it eroded personal ties of subordination, and gave rise to a notable process of land redistribution that survives to this day.[58] Sharecropping tended to disappear, small peasants could buy land and machin-ery and dream of becoming independent producers, middle farmers developed new commercial and productive strategies, and sons and daughters could lever-age their income against parental tutelage (for instance, around marriage strate-gies).[59] The process softened traditional class hierarchies and produced new class differentiations. Southern Catalans saw their world affected by novel processes beyond their control, nurturing feelings of victimization and anxiety.

The Struggle for Dignity

Victimization and anxiety did not translate into passivity. In the 1960s, new initiatives had already emerged in the Southern Catalan countryside. Indeed, the weakening of vertical ties opened the possibility for developing new forms of hor-izontal organization and, ultimately, for the politicization of the peasantry. This became most obvious in the recovery of the cooperativist spirit, with several vil-lages creating new municipal cooperatives in the 1960s and early 1970s. In 1964, Clara and Joan, both born in the 1930s, were deeply involved in the founding of a cooperative in Serra, a small village in the county of Priorat where they have lived all their lives.

The objective of the cooperative was, in their words, to "supersede the traders that were gouging us. We didn't want to do politics; we just wanted to live off the land, because that meant not having to leave." Yet this effort to build a space of autonomy that could improve their livelihoods had a broader transformative effect on their lives, as Joan explains:

> When we got married I thought that God was the most important thing in the world, that he decided it all. But with the co-op, as we started getting more involved, we got in touch with other people, anarchist and communist houses, those that had suffered the repression, and you start seeing things differently, you see the injustice of it all, you see that justice is in the hands of the opportunists [*aprofitats*], not the hard-working people, and there is a moment in which going to Easter procession does not mean anything to you. I am starting to cry.... You see that it's all a lie. You awake: the Church is not the only thing, you see there is another way to understand life, that you can change it.

Later on, in the late 1960s, Clara and Joan decided to go a step further. They joined efforts with four other families, all active members of the cooperative, and dug a deep well in their land in order to increase production. They became the first farmers to go organic in the region.

But all these initiatives were an economic failure: "In trying to find an economic solution, we became environmentalists and peasant activists. But nothing worked. Slaves were tied in chains, and they hurt. We [the peasants] are so well tied that it doesn't even hurt. They have massacred us. But we must be hopeful, otherwise God will be back." This quote illustrates the frustrating position in which the Southern Catalan peasantry found itself during the Miracle: despite increased access to the means of production, their new relationship of dependence on the market meant that they could hardly eke out a living. They were "squeezed." Yet the quote also makes clear that the dissolution of the traditional agrarian order revitalized the secular search for autonomy.

The radical peasant union Unió de Pagesos (UP), created in 1974 as a clandestine organization, would give organized expression and political direction to these hopes and anxieties of the Southern Catalan peasantry. Clara and Joan joined it from its inception, as did hundreds of middle and small peasants trying to change their lives. The formidable capacity of UP to become the leading voice of the Southern Catalan peasantry right from the moment of its foundation, when chapters were established in almost all Southern Catalan villages, cannot be understood without considering the political effervescence of the waning years of the dictatorship. UP's first front of struggle was the vertical farming organizations created by the regime and controlled by local cronies; to destroy them was to eliminate a local bastion of the regime's power. Furthermore, local

UP chapters incubated the first left-wing municipal governments elected in the late 1970s (the first democratic elections in four decades).

UP was a major actor in the democratization of the Southern Catalan countryside, connecting the region to the political struggles taking place throughout the country. Yet UP understood the fight for a new democratic regime as part of a larger, older struggle. Indeed, its founding manifesto proclaimed UP to be the heir of Unió de Rabassaires, the old sharecropper union. That same manifesto situated the plight of the peasantry in a long historical context of exploitation, dependence, and marginalization: "The Catalan peasantry finds itself in a critical, in some cases desperate, situation. The analysis of the current conjuncture clearly manifests that we are victims of a process of capitalist accumulation that has been built on our backs, sucking from the Catalan countryside not only our savings ... but also the best part of our labor power: the youth that has had to emigrate to increase the army of industrial workers.... During [the dictatorship] the peasantry has been the most ravaged sector within the whole economic system."[60]

UP's strength stemmed from its ability to articulate the secular struggle against relations of dependence in a historical context where the nature of this dependence was shifting. The old peasant claim of the *rabassaires* "Land for the tiller" morphed into "Being able to live off the land" (*Volem viure de la terra*). This neatly captured the transformation of the Southern Catalan rural world: in the 1970s, sharecroppers, agrarian wage laborers, and landless peasants were nearly extinct.[61] However, at the same time that the peasantry had accessed the means of production, especially land, it did not possess its means of reproduction. Peasants were unable to make a living from the land. If in the late-nineteenth century Lenin believed that traditional agrarian ties of personal dependency led to an "abuse of the human dignity," UP argued that market dependence eroded the bases of farmers' livelihood, "maintain[ing] peasants and nonpeasants in indignity."[62] It is in this historical moment that the Southern Catalan secular struggle for autonomy adopted, for the first time, the idiom of dignity.

Indeed, UP's fight against peasants' "indignity" was a struggle against "dependence," which operated in two interlocking forms: subordinating the peasants to the interests of capital, thus making it impossible for them to make an autonomous living; and subordinating the country to the city, thus maintaining the rural dweller as a second-class citizen. These two ideas find clear reflection in UP's two primary demands. One was a demand for provision of social services and amenities in rural areas (schools, hospitals, cultural centers, agrarian institutes) to counter the exodus of population and the stigma of backwardness.

Second, UP understood peasants as "'workers of the land' [*trabajadores de la tierra*] demanding a just compensation for their work and means of production fundamentally through 'fair prices' for their production."[63] It is in this second aspect that UP's activity was most visible, particularly as it undertook a series of direct-action strategies, most notably road blockades. While presenting the peasants as workers showed a clear desire to situate their plight in terms of political economic structures and within a broader context of working class struggle, UP always emphasized the term *pagès*, and it did so in a context in which the *pagès* was supposed to either disappear or mutate into a modern farmer. But UP desired to remain attached to family farming, to preserve certain traditional dynamics that provided some autonomy, and to protect the reproductive logics of the small producer from the ravages of the market.

If we look at the language and actions of UP, we may notice that they seem to inhabit different historical times and political projects. On the one hand, we have workers, exploitation, and democracy, words that seem to capture the spirit of the times; on the other, we have peasants, fair prices, and dignity, words with moral undertones, apparently alien to the historical context and political economic struggle. This apparent mismatch can be squared by understanding the struggle of the Southern Catalan peasantry as an instantiation of what Ernst Bloch called "nonsynchronous" movements.[64]

For Bloch, nonsynchronism should not be confused with anachronism or backwardness. Nonsynchronous contradictions are such not because the objective conditions that lie at their root are noncontemporary or because the groups that experience them are "out of sync." Rather, nonsynchronous movements emerge when a group does not engage with those contradictions in terms of the temporality and the forms of experience and consciousness dictated by state and capital. The idiom of dignity emerged to point at that nonsynchronism. Indeed, the very characterization of Southern Catalans as semi-proletarians misses something. In presenting their efforts to make a living as a simple expression of the capitalist dynamics of production—of the requirements of accumulation— Southern Catalans' long trajectory of struggle to reproduce their livelihoods and escape relations of dependence is overlooked.[65]

The struggle of the Southern Catalan peasantry represented the emergence into public life of an unsettled memory inscribed in contemporary practices and values and heralded by a population repeatedly excluded from the march of modernity and the body politic. It was a nonsynchronous struggle because it could only emerge by disrupting a teleological narrative of modernity and economic growth that condemned them to irrelevance and dispossession. Yet inseparable from the nonsynchronism that points toward an *uncompleted past*, this struggle also possessed a synchronous dimension signaling beyond an *impeded future* where the division between country and city could be superseded, peasant livelihoods improved, and relations of dependence recognized and challenged. This is what the demand of dignity of the Southern Catalan peasantry was all about: dignity against degradation and worthlessness, dignity against passivity and dependence.

The struggle for dignity re-emerged in a historical moment in which, in a new twist of the division between country and city, Southern Catalonia would be assigned a new function. In the mid-1970s, migration from the country to the city slowed dramatically, a consequence both of the economic problems unleashed by the oil crisis of 1973 and the exhaustion of the pool of rural labor. The grueling situation of family farming and the end of the exodus signaled the moment when rural regions such as Southern Catalonia would effectively become redundant, places—to borrow Gavin Smith's expression—"with no productive function" within Spain's structure of accumulation.[66] In that moment, the region was assigned a new economic function as an energy-producing hub. This is how Southern Catalonia would be subjected to "improvement," that is to say, enhanced for profits and capital accumulation.[67] Profits and energy would be extracted, sent to industrial and financial centers. Southern Catalonia would consequently be transformed into a new kind of periphery: its inhabitants and their activities would remain redundant and symbolically marginalized, while its territory would emerge as a desirable resource for capitalist accumulation and extraction.[68] The process had started timidly during the 1960s, with the construction of a handful of large hydroelectric dams. It would expand dramatically in the 1970s and 1980s with the installation of the nuclear plants. Energy extraction was what modernity would finally bring to Southern Catalonia. It would come as no surprise that the peasantry, in their struggle for dignity, would fight against it.

Notes

1. Marx and Engels (1986: 25).
2. This was the expression officially adopted by the regime (Sevilla Guzmán 1979), and it has subsequently been amply used in the academic world. Despite the problems of the adjective "traditional," I have opted to retain the expression for the sake of clarity.
3. See Narotzky (1997: 216–218).
4. For a summary of the peasant dialectic between dependence and autonomy, see Narotzky (2016c); on the integration between peasant households and capitalist production, see Banaji (2016).
5. See Vilar (1964–68, especially volume 3).
6. For an overview, see Mayayo (1995) and Garrabou (2006).
7. On the Southern Catalan *casa*, see Jociles (1989) and Ferrús (1985). On "traditional" domestic structures and strategies in agrarian Spain, see Contreras (1991a).
8. On traditional hunting practices in Southern Catalonia, see Fucho (1998).
9. On the expulsion of household members in agrarian Europe, see Bourdieu (2008).
10. Costa (1901: 8–10).
11. Quoted in Domènech (2004: 421).
12. See Narotzky and Smith (2005) for a detailed discussion of these two strategies.
13. On the system of loyalties between sharecropper and landlord, see Contreras (1991b).

14. Garrabou and Saguer (2006). The Catalan case strongly contrasts with the situation in other parts of Spain, especially the south.

15. Godelier (2001, especially 19–20).

16. Terradas (1984: 268).

17. See Mayayo (1995) for a detailed description of these forms of struggle in the Catalan countryside; for the specific case of Southern Catalonia, see Sánchez Cervelló (2001).

18. The political consequences and class impulse behind the establishment of cooperatives was varied and often ambiguous (Planas and Garrido 2006). Nonetheless, in Southern Catalonia cooperativism was overwhelmingly aligned with the workers' movement (Audí Ferrer 2010).

19. Izquierdo Martín and Sánchez León (2010: 149).

20. Bloch (1991).

21. Quoted in Nel·lo (1991: 81).

22. On these two different notions of freedom, see Bloch (1986: 160); see also Wood (2015: 126ff). On the connection between republican citizenship and the abolishing of the city-country division, see Ross (2015: 85–90).

23. On this virulent opposition, see Mayayo (1995) and Tébar (2006). For a general overview of the Spanish republic and the Civil War, see Casanova and Gil Andrés (2014: 105–215).

24. Tébar (2006: 582).

25. For contrasting experiences in the region, see Narotzky (2007), Bailey (1990), and Simoni and Simoni (1984).

26. See Termes (2008) and Pagès (2004).

27. For a description of this battle's impact on Southern Catalonia, see Castell et al (1999).

28. Recasens Llort (2005); see also Solé Sabaté (2003).

29. For succinct, up-to-date descriptions of these two phases, focused, respectively, on their political and economic dimensions, see Casanova and Gil Andrés (2014: 217–287) and Carreras and Tafunell (2010: 263–366).

30. Sevilla Guzmán (1979: 141).

31. Quoted in Tébar (2006: 595).

32. Bretón (2000).

33. Carreras and Tafunell (2010: 276).

34. On the activity of the *maquis* in Southern Catalonia, see Sánchez Cervelló (2003).

35. Contreras (1991b: 514).

36. Carreras and Tafunell (2010: 281).

37. On the role of Western powers, see Garcés (2012).

38. More than 2 million Spaniards emigrated to Europe between 1960 and 1977 (Martínez Veiga 1991: 233).

39. On the role of the US and the World Bank in the design of the policies of *desarrollismo*, see Muñoz et al. (1979) and Larrú (2009).

40. For a theoretical discussion of the extension and limits of this project, see Smith (2011) and Collins' (2011) response.

41. Quoted in Sevilla Guzmán (1979: 205).

42. Old areas of irrigated agriculture maintained their population in absolute—but not relative—terms (Clar and Silvestre 2008).

43. Bretón (2000: 178) and Majoral (2006: 615). It should nonetheless be noted that the composition of the labor force was very different in Catalonia and the rest of Spain. In the

former, the industrial workforce overcame the agrarian one by 1910, something that would not occur for the whole of Spain until the 1960s (Mayayo 1995: 23).

44. Marx (1976: 472; emphasis added).

45. Gramsci (1957) and Williams (1973). See Brennan (2017) for a recent discussion.

46. The numbers for the whole of Spain show a remarkable trend: from 56,800 to 355,000 tractors between 1960 and 1974; from 36.9 to 87.8 kg of fertilizer per cultivated hectare (Palomera 2015: 19). A new energetic balance governed the relationship between the two sectors: in 1950, for every calorie invested in agriculture, there was a return of six calories; by 1980, the figure was a meager 0.7 (Naredo 1986).

47. Perelman (2000).

48. During the postwar period, agricultural prices augmented at a higher rate than salaries. This tendency started to invert between 1953 and 1958 (Naredo 1986), continuing to this day: in the year 2000, agrarian revenues supposed only 55 percent of the total revenue of farming households (Etxezarreta 2006).

49. On this discourse, see Sevilla Guzmán (1979: 203–242).

50. Quoted in Bretón (2000: 190). The quote is from Martínez Borque, director of the Instituto Nacional de Colonización (INC).

51. Terra Alta and Priorat lost as much as 30 percent of their population between 1950 and 1975 (Grau Folch 1993, Margalef and Tasias 1985). In Ribera, the decline is more moderate—12 percent between 1965 and 1975 (Sorribes and Grau Folch 1989).

52. This is a clear instance of what Trotsky (1906) called "combined and uneven development."

53. See Cucó (1982) for an ethnographic analysis of this process in the Valencia region.

54. This process is known as "peasant squeeze"; see Bernstein (2010).

55. Narotzky (1997: 191–203).

56. Etxezarreta (2006); Contreras (1991a).

57. Soronellas (2006).

58. On the concentration of agrarian property in different Spanish regions, see Soler and Fernández (2015: 32–36).

59. In the late 1980s, sharecropping contracts represented less than 1 percent of farmed land (Grau Folch 1993: 73).

60. Quoted in Peix (1999: 25–26).

61. This was only the case in the areas of dryland farming; in the irrigated areas, especially abundant in the south of the region, agrarian wage labor remained numerically very important.

62. Lenin (1978: Part III); the UP manifesto was part of the conclusions of the Committee on Agriculture at the Conference of Catalan Culture, celebrated between 1975 and 1977; accessed October 3, 2017, http://congres.fundccc.cat/ambits/ambit-dagricultura/resolucions-de-lambit-dagricultura/.

63. Etxezarreta (2006: 295).

64. Bloch (1991, especially 97–116).

65. On the peasants as political subjects, see the classic studies of Moore (1966) and Wolf (1969).

66. Smith (2011).

67. On the early usage and etymology of "improvement," see Wood (2002: 106–111).

68. Despite other dissimilarities, this situation is structurally analogous to the one described by Li (2010).

2 Nuclear Transaction

I would like to call the whole Spanish people to the task of overcoming the night in energy matters. Not to immediately implement the National Energy Plan, with all its fundamental options, would purely and simply be betting on obscurity [tinieblas] and getting ready to govern in the dark [a oscuras] by 1987.

—Agustín Rodríguez Sahagún, Minister of Industry and Energy (1978–1979)[1]

FOR MOST OF the twentieth century, nuclear energy was seen as the beginning of a definitive energy transition, a technological achievement that would emancipate humankind from resource dependence, scarcity, toil, and conflict. The following quote by R.M. Langer, a physicist at the California Institute of Technology, summarizes this vision: "The face of the earth will be changed.... Privilege and class distinctions will become relics because things that make up the good life will be so abundant and inexpensive. War will become obsolete because of the disappearance of those economic stresses that immemorially have caused it.... The kind of civilization we might expect is so different in kind from anything we know that even guesses about it are futile."[2]

Although the military use of nuclear energy put a check on this utopian vision, in the 1960s and 1970s, nuclear energy was still seen as the basis for "*definitively* solving the energy supply problems of industrialized countries" and ridding them of their dependence on exhaustible fossil resources.[3]

Indeed, in the early 1970s, the Spanish government and the main electricity corporations of the country joined efforts to develop an ambitious nuclear program involving the construction of up to thirty-five nuclear reactors and the full-fledged nuclearization of the energy supply. Southern Catalonia was to be extraordinarily affected by this process, becoming the most nuclearized region in Spain. The program for a nuclear transition was in large part, so said its advocates, a response to the oil shock of 1973 and the ensuing energy crisis. In this context, proponents hailed the transition to nuclear energy as a Promethean cornucopia: Promethean inasmuch as it represented the ultimate victory of human inventiveness over natural scarcity, finally unshackling humanity from the constraints of material finitude; cornucopian inasmuch as it would rid the country of its dependency on foreign energy resources and allow it to keep pace with its economic development. Nuclear dreams were finally destroyed in 1983, when the Spanish government decreed a nuclear moratorium—technically still in place—that put an abrupt end to a nuclear program that had become increasingly unpopular and had brought the Spanish electricity sector to the brink of financial collapse.

The timing of the failed transition to nuclear power is highly relevant and revealing, for it largely coincided with the *Transición*, the decade-long (1973–1982) liminal political period during which Spain transformed from a dictatorial regime into a liberal democratic kingdom. The objective of this chapter is to analyze the relationship between these two transitions—between an ultimately failed transition to nuclear power and a seemingly exemplary transition to democratic rule. Critics often describe the truncated nature of the *Transición*, playing with the words *transition* and *transaction* (meaning a "deal") to suggest that the *Transición* was not so much a true *transition* involving a clear-cut change of political regime and the construction of new socioeconomic structures, as a *transaction*, a deal between old and new political elites to preserve the socioeconomic status quo. My thesis is that the nuclear program was a crucial element operationalizing this transaction. On the one hand, it facilitated a top-down management of the political process that limited its democratic openness and potential for rupture. On the other hand, it offered the ruling economic elites articulated around the electric sector the opportunity for economic accumulation and expansion in a context characterized by political and economic uncertainty.

Thus, in order to understand the nuclear program and its consequences, we must first historically examine the creation of Spain's electricity sector.

Water and Spanish Modernity

In the 1970s, the advocates of nuclear power presented its development as a necessary element for the modernization of Spain. Yet up until that historical moment the projects and dreams to modernize the country had revolved around a different energy source: waterpower. Water first emerged in the national debate in the late nineteenth century through the works of a diffuse group of intellectuals known as *regeneracionistas*. For these intellectuals, the modernization of Spain could be achieved through an ambitious series of hydraulic projects extending irrigation. The remaking of Spain's waterscape would multiply agrarian production, creating a new mass of agrarian smallholders and doing away with *caciquismo*. Together with modernizing engineers, enlightened intellectuals, and industrialists, this mass of agrarian petty owners were to constitute a hegemonic bloc that would lead Spain to find a third way between the existing oligarchic structure of the country, dominated by rentier classes, and the threat of proletarian revolution.[4]

Water politics and the construction of hydraulic infrastructure were to play a crucial role in Spain's twentieth century. Yet this role was to be very different from the one that the *regeneracionistas* had envisaged: dams and reservoirs were not to be primarily devoted to irrigation, but to the production of electricity; water politics was not to encourage the democratization of Spain, but rather reinforced dictatorship; and instead of leading to the formation of a new hegemonic bloc, the remaking of Spain's waterscape was instrumental in consolidating the power of oligarchic elites.

Capitalist Dreams, Monopolistic Projects

Electricity expanded quickly and from early on in Spain: by 1890 there were as many as a thousand plants devoted to electricity generation.[5] From 1880 and up until 1910, the electricity system in Spain (and elsewhere) was dispersed and fragmented, largely mirroring settlement and productive patterns. Indeed, the expansion of electricity was especially noticeable in the country's two most industrialized regions: the Basque country—where heavy industry developed in the closing decades of the century—and Catalonia—where textile manufacturing had led the extension of a notable process of industrialization concentrated in the center and east of the region. Early electricity networks developed both in urban centers and in the countryside.[6] The former were powered by coal power plants, whereas the latter took their energy from water, often emerging around industries located by the river, which in Catalonia were especially numerous.[7] Most of the companies running these systems were very small, with the only exception of those that served large urban markets, such as the Compañía Barcelonesa de Electricidad (CBA), created in 1894 by the German electric giant AEG. Yet in all cases, the technical difficulties of electricity transmission involved that electricity networks were bound to the local scale.

As Thomas Hughes showed in his classic book, *Networks of Power*, electricity systems all over Europe and the Americas experienced a dramatic transformation in the first third of the twentieth century, evolving in scale from small lighting systems extending just a few city blocks to complex regional generation systems combining diverse power sources with extensive distribution grids operated with sophisticated managerial methods. Hughes explains this transformation as the result of a double process of *concentration*, led by technological innovation, and *centralization*, spurred by intercapitalist competition. The improvements in electricity transmission made possible the concentration of electricity production in increasingly large power plants (mostly, hydroelectric dams) supplying distantly located urban-industrial centers. In turn, this made possible a techno-managerial innovation in *load management*: the electricity produced through myriad technologies and resources was managed as a single load of bulk power, an undifferentiated magnitude to be maximized by adjusting it to the variegated end uses of electricity—private lightning, industrial power, urban transportation systems. In so doing, large electricity utilities were able to create economies of scale that expulsed smaller, local providers from the market, creating the conditions for the centralized control of the electric system in fewer hands.[8]

In Spain, the origin of the complex socio-technical process that Hughes describes can be quite accurately dated to a meeting in Barcelona in 1911 between Frank Stark Pearson and Carlos Montañés.[9] Montañés, a Catalan engineer with close links to the city's thriving bourgeoisie and deputy in the Spanish parliament, had long cherished the dream of electrifying the industrial fabric of Barcelona's urban region. Yet he lacked the financial resources to make it a reality until a good friend, a Spanish diplomat who was a board member of a major UK bank, put him in contact with Pearson, a US-born electricity tycoon with extensive interests in Brazil and Mexico. In 1911, exactly at the moment that his Latin-American businesses were under stress and scrutiny, Montañés would offer Pearson his last frontier, and he would do it from the top of Tibidabo, a small mountain on the edge of Barcelona. Its name means "I will give you."

Pearson went to Barcelona to meet Montañés. Early the next morning, the Spanish engineer picked up the American from the hotel, saying he "wanted to show [him] a panoramic vision of our project's objectives."[10] During the car ride, Pearson told Montañés that he was very skeptical: his experience in Latin America had made him reticent about doing businesses in Latin countries. Secure in his skills and contacts, Montañés reassured him: "I will personally take care of all these matters: everything related to political intromission, the administration and the public opinion."[11]

At the top of Tibidabo, Montañés spread his maps and described the details of the project. Then, the vision that he wanted Pearson to admire started to take place: "The chimneys of the factories of Barcelona [and its surroundings] started to exhale abundant smoke.... That was the best evidence of the power installed in that area."[12] Montañés recalled that Pearson "grabbed my shoulder, looked at me for a few seconds, and said: 'Montañés, jewels like this one barely exist in the world anymore ... I'll take the business.'"[13]

Pearson's words sealed the deal for the constitution of the Barcelona Traction, Light and Power, headquartered in Toronto.[14] The company built a comprehensive system of electrification in Catalonia, the first regional network in Spain. Yet the realization of this project crucially relied on two operations that could not be observed from the top of Tibidabo. First, eliminating a whole host of smaller electricity utilities; second, producing large quantities of power. The achievement of the first relied not only on market mechanisms but also on Montañés's political contacts and Pearson's knowledge of the high spheres of capitalism, including an unannounced trip to Berlin to broker a personal deal with AEG's CEO, Emil Rathenau, to purchase CBA. By the early 1920s, Barcelona Traction had achieved a semi-monopolistic position in Catalonia, and only one smaller company, Hidruña, resisted its push.

To achieve the second objective, Montañés and Pearson built a system of hydroelectric dams in the Catalan and Aragonese Pyrenees. Once Pearson's team of engineers started building these dams, Montañés recalls observing "the American spirit of 'New Frontier,' the 'go west young man, go west' of the nineteenth century, which in our country, for those technicians who mastered the most advanced technology, translated into perforating mountains and wading across abysses."[15] For half a century the "white coal rush" continued in the Pyrenees, converted into the main source of power consumed by Catalan towns and industries. In the meantime, in the area of activity of the Barcelona Traction alone, two hundred small hydroelectric dams owned by industries and municipalities to satisfy local demand—the remnants of the old, fragmented electricity system—were rendered obsolete.[16]

The centralization and concentration of the electricity system promoted the centralization and concentration of capital, population, and economic activity. Indeed, the new hydraulic landscape created by companies like the Barcelona Traction did not strengthen the bonds between agrarian and industrial productive classes, but rather reasserted the division between country and city, as Pierre Vilar noted: "Born for Barcelona and thanks to it, [the Barcelona Traction's] hydroelectric network did not bring anything new, either to the mountain or along the [electric] transportation lines. Ninety percent of its energy was consumed by Barcelona and its surroundings."[17] Already by 1921, the Barcelona Traction was not only the largest Spanish electricity company; it was the largest corporation in Spain.[18] But it was not the only electricity company in the top tier of Spain's corporate hierarchy. At the dawn of the Civil War, it was joined by a series of other electric utilities that had also established regional monopolies in other Spanish regions, largely corresponding to the major river basins.[19] They constitute the kernel of Spain's current electricity sector.

Fascist Dreams, Corporatist Power

With the war and the subsequent repression, Franco eliminated oppositional movements and subdued labor, while securing the loyalty of a heterogeneous mix of royalists, fascists, Christian fundamentalists, the church hierarchy, and a sizable portion of the industrial bourgeoisie, including the electric utilities. The linkages between the state and the economy were strengthened, consolidating a corporatist state structure that relied on an endogenous capitalist sector whose profits were closely tied up with the state's investment flows. Hydroelectricity played a critical part in this process.

Immediately after the war, there was a heated debate within the Francoist ranks over the future of the electric system. Fascist ideologues proposed to nationalize the electric utilities. The owners of the utilities, who possessed deep connections with other sectors such as banking, steel, and transportation, resisted the idea. The conflict was resolved through a personal deal between the dictator and the ultra-Catholic president of Hidrola—later Iberdrola—José Maria de Oriol.[20] The resulting compromise established a system controlled by about a dozen corporations with regional monopolies. This group of companies is to this day popularly known in Spain as the "electrical oligopoly" or simply the "oligopoly." Although the system was nominally self-regulated, the regime, which never withdrew the threat of nationalization, exercised tight control over it by regulating prices, establishing profit rates, and targeting strategic investment. The regime, for instance, obliged the utilities to build interconnections between their networks, leading to a new jump in scale in the electricity system, which encompassed the whole country. In addition, the government created two powerful public electricity corporations, ENDESA (Empresa Nacional de Electricidad) and ENHER (Empresa Nacional Hidroeléctrica de la Ribagorzana).[21]

No event captures the corporatist character of Franco's regime as the 1952 takeover of the Barcelona Traction, the only major electricity company not controlled by Spanish capitals. Through a controversial judicial decision that shook the bases of international law, the Spanish courts declared the company bankrupt, transferring its ownership to a series of bondholders led by Juan March, a Majorcan businessman and contrabandist that had funded the military uprising of 1936. With the takeover, the Barcelona Traction was renamed Fuerzas Eléctricas de Cataluña (FECSA). Bankers and representatives of other utilities sat with former ministers and high-ranking military on its first board, evidencing the bundle of power interests coalescing around the electricity sector.[22]

The centrality of the state in organizing the electricity system became most obvious in the construction of hydroelectric dams, a process led by the state-owned companies ENDESA and ENHER. The construction of dams and reservoirs was a central piece of the regime's propaganda. With these hydraulic works, the regime appropriated the modernizing, populist rhetoric of the *regeneracionistas*, as it can be appreciated in a speech that Franco pronounced in Lleida in 1959: "We have come to visit your province, to inaugurate various important works ... and with this to satisfy the thirst of your fields, to regulate your irrigations, which shall increase your welfare and multiply production.... The whole of Spain needs to be redeemed, sealing the *brotherhood* between the land and the men of Spain."[23] Yet despite this rhetoric, the main objective of these projects was

Table 2.1. Electricity production in Spain by primary source, 1931–1995 (in GWh).

Year	Hydro	% Total	Fossil Fuels	% Total	Nuclear	% Total	TOTAL
1931	2,381	88.8	300	11.2	—	—	2,681
1935	2,993	91.5	279	8.5	—	—	3,272
1940	3,353	88.5	264	11.5	—	—	2,290
1945	3,180	76.2	993	23.8	—	—	4,173
1950	5,079	73.5	1,837	26.5	—	—	6,916
1955	8,991	75.4	2,931	24.6	—	—	11,922
1960	15,624	83.9	2,990	16.1	—	—	18,614
1965	19,686	62.1	12,037	37.9	—	—	31,723
1970	27,959	49.5	27,608	48.9	923	1.6	56,490
1975	26,474	32.1	48,489	58.8	7,543	9.1	82,481
1980	30,807	27.9	74,490	67.4	5,186	4.7	110,483
1985	33,033	25.9	66,285	52.1	28,044	22.0	127,363
1990	26,183	17.2	71,289	47.0	54,267	35.8	151,740
1995[1]	24,203	14.8	76,320	46.7	55,445	34.0	163,308

SOURCES: 1931–1965: *Anuario Nacional de Estadística* (www.ine.es); 1970–1990: *Estadística de la Energía Eléctrica* (www.minetad.gob.es); 1995: adapted from *El Sistema Eléctrico Español* (www.ree.es).

1. Total does not add up to 100 percent. The missing 4.5 percent corresponds to cogeneration (based on fuel-oil and natural gas).

not irrigation but the production of electricity. As it can be observed in table 2.1, hydroelectrical generation increased exponentially, especially after 1955 with the beginning of economic liberalization.[24]

It is in this moment that Southern Catalonia first entered the picture. Indeed, between 1958 and 1967 the state-owned company ENHER promoted the construction of the dams of Mequinensa and Riba-roja, in the Aragonese lands adjacent to Southern Catalonia. These projects flooded fields and entire villages, whose inhabitants were relocated in newly built towns. One telling fact captures the despair of this population: within the first year of the relocation, the number of deaths of old people multiplied fivefold.[25] The impact of these dams on Southern Catalonia was less dramatic, but also noticeable: they interrupted the flow of sediments and fluvial navigation, thus endangering the Ebro Delta and destroying one of the main sources of economic activity of riverine villages. But it could have been much worse, for Hidruña (the second largest Catalan electricity company) projected a third dam, located in the town of Xerta, which was supposed to flood the most fertile, irrigated lands of the county of Ribera. The local population reacted swiftly, launching a judicial campaign that pushed Hidruña to temporarily withdraw the project in 1964.[26] From that moment onward, every new energy project in the region would elicit a desperate yet robust local opposition campaign.

Developmentalist Dreams and the Brotherhood of Cement

Hydroelectric projects played a crucial role in organizing the structure of accumulation of the Miracle. On the one hand, they were the quintessential expression of the state's effort to create the necessary infrastructure for capital accumulation, the state-controlled abstract space that capital required.[27] This is what the modernization process that took place during *desarrollismo* was all about. On the other hand, they created an electricity system that functioned as a crucial arena for the coordination of state and capital interests. The electric sector was deeply intertwined with the political sphere, and came to represent, along with the banks, the political-cum-economic power nucleus of the dictatorship. The construction sector, born around the state's developmentalist projects, was the third leg of the inner core of the corporatist state. Let's observe how these three sectors articulated around the electricity system.

In the 1970s, the electric sector was the principal center of attraction and centralization of capital within the Spanish economy. In 1975, the twelve companies that composed the oligopoly were all among the thirty largest corporations in Spain, absorbing 70 percent of all bonds.[28] The centrality of the electricity corporations within the Spanish economy was never based on their profits, but rather on the control of vast financial resources and a huge market that could be used to generate profits and monopoly rents in other sectors such as construction, banking, engineering, electrical materials, and so forth. The electric sector had no internal or external competition. Internal competition was limited by the electrical companies' quasimonopolistic control over certain geographical areas and the regulated price of electricity. It was, at the same time, a very cohesive sector and with a high degree of interpenetration, evident, for instance, in the joint ownership of the electricity grid. The entry of new actors was severely limited by the regime's blockage of foreign investment in a sector deemed "strategic," a consideration that also extended to the banking sector.[29] Indeed, by protecting the sector from the liberalizing dynamics that characterized *desarrollismo*, the regime reinforced the oligopoly's position as a bastion for the country's political and economic elites.

The major Spanish banks, the real centers of decision making within Spanish capitalism, controlled the electrical sector and occupied most of the seats on the sector's boards. We must also note that the banks had control of these companies without having majority ownership of them, a circumstance made possible by their fragmented structure of stockholding, with a strong presence of small stockholders—up to a million stockholders for the whole sector in

1977—that reflected the degree to which these companies absorbed Spanish savings.[30] Finally, the creation of public infrastructure, and especially the building of hydroelectric projects, provided the perfect context for the emergence of a strong construction sector benefitting political cronies. Indeed, the vast majority of the construction companies building state-funded hydraulic projects had been created in the immediate postwar period.[31] Private utilities also constituted their own construction companies to build their hydraulic infrastructure. Both kinds of construction companies soon expanded beyond the energy sector, building the roads, canals, airports, hotels, and housing complexes that the new economy required.

Far from the alliance between entrepreneurial farmers and liberal middle classes envisaged by the *regeneracionistas*—or the brotherhood between "men and land" of Francoist propaganda—the regime's hydraulic projects were instrumental in creating a hegemonic bloc constituted by the successors of the old agrarian, rentier, and financial sectors. Together with state engineers, these three sectors—banking, utilities, and construction—formed what Martínez Gil calls the "brotherhood of cement."[32] The brotherhood of cement equated modernization with the construction of large-scale projects, thus conflating the supposedly neutral character of development with its own interests. Thus, for instance, electric utilities were majority owners of most cement factories, therefore having a strong incentive to promote new facilities, built through construction companies that were controlled by the same banks that sat on their boards. The nuclear program would extend the connection between construction, banking and the electricity sectors to the limit.

The Nuclear Program

The nuclear program was the culmination and epitaph of Francoist *desarrollismo*. *Desarrollismo* had a strong impact on the patterns of energy production and consumption in the country. First, rapid industrialization led to a large increase in demand for electricity, with annual rates of increase—approximately 10 percent—that were, in fact, higher than the GDP growth rate, a circumstance explained by the privileged role that the economic development strategies of the government reserved for high-consuming industries such as cement, steel, aluminum, and (petro)chemicals.[33] Second, despite the monumental hydraulic projects built for Franco's glory, the late 1960s were fundamentally marked by the rise of Middle Eastern oil as the main energy source in Spain, which represented 20 percent of total energy consumption in 1955 and 65 percent in 1973.[34] The increased use of oil, an energy source denominated in US dollars, was certainly linked to the increasing influence that the United States had over the Spanish economy.[35] The increased importance of oil also affected the electric sector. Indeed, during the 1960s, while the state-owned electric companies were busy building hydroelectric dams, the private utilities engaged in an aggressive plan to install thermal power plants functioning with fuel oil. As a result, by 1970, fossil fuels had overtaken waterpower as the main source of electricity production (see table 2.1).

Third, the early adoption of nuclear energy in Spain was also linked to the good relations with the United States and that country's strategy of promoting the civilian use of nuclear power among its allies, first announced internationally in 1953 with Eisenhower's "Atoms for Peace" speech to the United Nations. Although in the early 1950s the Spanish government treated the nuclear question as a military matter, by 1956 it was already encouraging the oligopolistic electricity corporations to explore the possibilities of nuclear energy.[36] In 1962 these companies created the Atomic Forum, a lobby to promote the private development of nuclear energy. In 1968, Spain opened its first nuclear power plant near Madrid, with an inauguration that included a blessing by the bishop of the country's capital, which thus made clear that nuclear energy represented the might of the nation.[37] Two more nuclear reactors, one in the Southern Catalan village of Vandellòs, completed what is known as the first generation of Spanish nuclear power plants.

The crucial turn to nuclear energy took place in the 1970s with the second generation of nuclear reactors. This second generation used US nuclear technology (from General Electric and Westinghouse) and relied on massive credits in US dollars obtained through the Export-Import Bank of the United States and underwritten by the Spanish state.[38] In 1975, just a few months before the death of the dictator, the Cortes (the legislative body of the Francoist regime) passed the first Plan Energético Nacional (PEN, National Energy Plan). Putting a strong emphasis on energy sovereignty and security and assuming the Spanish economy's steady growth in demand for energy, the PEN of 1975 (PEN-75) was intended to reduce the country's dependence on oil. In practice, this goal was to be achieved through the electrification of energy supply and the formidable increase in the production of nuclear energy. The 9.1 percent of electricity that had a nuclear origin in 1975 was supposed to grow to 56 percent by 1985, when, according to the PEN-75, Spain would have thirty-five nuclear reactors in operation.[39] Of these, seven were to be located in Southern Catalonia, which would thus become one of the most heavily nuclearized regions in Europe. The scale of the project was fabulous: the planned financial investment of the electricity corporations would double the fixed assets that they owned.[40]

Indeed, the nuclear program was the dream of the brotherhood of cement, extending the connection between banking, construction companies, and the electricity sector to the limit. In the context of the industrial stagnation and political turmoil that characterized Spain in the 1970s, the nuclear program emerged as a fantastic opportunity for the capitals controlling the electricity sector, which realized that extraordinary profits could be made from the construction of power plants rather than from the production of kilowatts.[41]

Indeed, the construction of nuclear power plants followed patterns common to large-scale projects whose logic has been independently theorized by Gustavo Lins Ribeiro, José Manuel Naredo, and Bent Flyvbjerg.[42] Each of these authors has analyzed large sets of what they call large-scale projects or megaprojects—large infrastructural works such as dams, bridges, airports, and irrigation projects— and observed recurrent patterns that can be grouped into three categories. First, these projects are presented as urgent and as "the best and most reasonable solution for an economic or social problem of national or regional scope," mere technical solutions that do not need to be open to political discussion.[43] As was the case with the Spanish nuclear program, these strongly ideological legitimations tend to rely on some notion of progress or development that presents them as good for everyone, thus obscuring their class dimension.

Second, by concentrating huge amounts of resources at a particular point in space, these projects are able to centralize, coordinate, and expand capital. Indeed, the sheer scale of the megaproject tends to exclude small capitals in favor of capitals operating in larger circuits, thus generating the opportunity, as Naredo argues, for the constitution of "oligarchic groups" able to use the apparatus of the state for their own benefit. This power dimension is also reflected at the local level, for the isolated locations of large-scale projects such as nuclear power plants favor the power of the developer over workers—often migrant workers— and over the surrounding regions, creating a controlled territory that tends to experience a cycle of boom and bust.

Third, as Flyvbjerg has consistently demonstrated, these projects almost invariably result in huge economic losses that end up being covered by the state and/or translated into losses for the small stockholders. These financial losses are usually caused by the brutal increase of construction prices, which operates as an extraordinary profit-making mechanism. The Spanish nuclear program, as we will see, was no exception to this rule.

A Revolt against Authority: Nuclear Projects in Southern Catalonia

To the surprise of political authorities and electricity corporations, the PEN-75 triggered antinuclear mobilizations all over Spain in the second half of the 1970s. In addition to having a strong influence in the rural areas where the nuclear plants had been planned, the antinuclear movement achieved a notable cohesion with urban sectors as well, constituting a highly political movement intimately connected both to the struggle against Franco's regime and to the emergence of modern Spanish environmentalism, especially its most political and combative current.[44] This struggle was decisive in forcing the government to cancel a large number of these plants, including three in Southern Catalonia. Ultimately, however, as Joaquín Fernández argues, this region "ended up taking the worst part, despite the outstanding popular response and some cases of heroism widely admired in the mythology of the ecologist movement."[45]

The nuclearization of Southern Catalonia spanned two decades and was experienced in the region as a moment of epochal change. It started in 1967, when a conglomerate of French and Catalan corporations using French technology started construction of a relatively small nuclear reactor in Vandellòs, a small village by the Mediterranean. Yet the main chapter took place in the 1970s when FECSA emerged as the leading force behind the nuclearization of Southern Catalonia. Indeed, the antinuclear movement coined the expression *fecsisme*, playing with the words *FECSA* and *fascism*, to oppose the construction of nuclear plants and to denounce the collusion between the electrical oligopoly and Franco's regime. In the spring of 1974, following the oil crisis of 1973, FECSA sent a report to stockholders announcing the "end of cheap energy" and the immediate arrival of a nuclear future for the company and the country.[46] Indeed, the international events of 1973 proved to be advantageous to a company that by then had already put all its eggs in the nuclear basket.[47] In the years that followed, this corporation, which assumed costly loans in US currency, would be involved, as owner or co-owner, in all of the nuclear projects underway in Southern Catalonia.

At the center of the antinuclear struggle was the village of Ascó, located fifty kilometers west of Vandellòs on the shores of another large body of water, the river Ebro.[48] Long before receiving the governmental authorization to build Ascó I in 1974, FECSA started to buy, secretly and for a very generous sum, the seventy hectares of land that would host the first of the two nuclear power plants that were to put the name of this village on the map and in the newspapers. The main local beneficiaries of the land purchase were the village large landowners, who had historically possessed the irrigated shores of the river, the most fertile land of the region. Indeed, the abundance of water needed to cool the nuclear reactor, a critical factor in a dry territory such as Mediterranean Spain, was one of the two main reasons FECSA chose Ascó for the location of the nuclear power plant. The project of a second plant, co-owned by FECSA and the public corporation ENHER, was announced and begun in the mid-1970s.

Early in 1970, once the secretive land deals had concluded, the daily press of Barcelona started airing the details of the "new factory in Ascó," a fabulous project that would provide two hundred jobs and require twenty-five hundred workers during its construction.[49] The first public acknowledgment of the project came a few weeks later, when the village mayor, a Francoist hard-liner, appeared on TV saying, "We are hopeful for the construction of the nuclear power plant, not only because it will contribute to the development of Spain, it will also solve the economic problems of the area, establishing the basis that will allow us to become an important industrial pole."[50]

Local accounts of this early period stress that FECSA told the villagers that the land purchase was for the construction of a chocolate factory. True or not, the tale of the chocolate factory, repeated throughout the region to this day, became a myth of origin, condensing the original sins of a facility that introduced an abrupt rupture in the life of the village and the whole region. The tale is critical of how the locals were treated, cheated like innocent children who accepted candy as bait. Even once the real objective of the land purchase was known to everyone, the tale of the chocolate factory worked as a powerful expression of the lies and misinformation on which the nuclear power plant was built, for the population was never informed of the dangers of a source of energy about which it was largely ignorant. The story also conveys a strong sense of unfulfilled promise: the inhabitants of Ascó were promised not only prosperity, but also a place in history, for they were repeatedly told that they were contributing to the might of the motherland. But in the end the motherland forgot about them, and the area did not become a pole of economic growth; perhaps more important, it was never meant to become one.

In fact, the second reason for locating the nuclear plant in Ascó was the low population density of the area—a legal requirement—and the low value attributed to its inhabitants. Social-democrat Josep Maria Triginer, president of the Commission for Industry and Energy of the Spanish Parliament, and by all accounts a strong ally of FECSA, cynically stated the point: "Many counties are not poor or depressed just because [*porque sí*], but due to a lack of investment. Yet other lands cannot be made profitable. So why do the nuclear plants get located in these areas? Well, we could say that while these power plants are much safer than other dangers that are part and parcel of our society, the consequences of a hypothetical accident are much more severe. This is why [nuclear power plants] must be situated in areas with low demographic density."[51]

According to this politician, Southern Catalonia's poverty was intrinsic, that is to say, it could not be remedied endogenously nor through exogenous investment. Thus, it was not to fix that situation that nuclear plants were to be installed there. On the contrary, it was because Southern Catalonia's intrinsic poverty made the region and its inhabitants dispensable that the area was suitable for hosting a series of plants that would supposedly yield fabulous profits.

This dispensable condition connects with an additional reason explaining the peripheral location of nuclear power plants all over Western Europe and North America. As George Caffentzis argues, the rationale behind this locational pattern was to avoid the labor struggles and increased politicization of urban centers by tapping into the "political passivity or conservatism of the local populations." Yet, he adds, "in this respect capital made some of the most painful miscalculations."[52]

Local ignorance about the nuclear danger is illustrated by the anecdote that triggered the organization of an antinuclear movement in the village of Ascó. In 1974, Andreu Carranza, a tailor with nine children and a devout Christian, had to accompany his wife, who needed to see a doctor, to Reus, the urban regional center. While he was in the waiting room, his attention was caught by a magazine containing an article with the title "The Threat of Nuclear Power."[53] Written by Mario Gaviria, pioneer of the environmental and the antinuclear movements in Spain, the article explained the risks of nuclear power, and it left Carranza very troubled. Back in Ascó, the tailor discussed the issue with Miquel Redorat, the new priest of the village, who shared the same concerns with his parishioner. From that moment on, the priest and the tailor formed a duo that became the vanguard of the antinuclear struggle in the village. Redorat was part of a group of radical priests, particularly well represented in Southern Catalonia, which would come to play a decisive role, especially in the early stages of antinuclear mobilization in the area. He used the pulpit to share weekly antinuclear admonitions, warning that nuclear power was dangerous for local livelihoods. His incendiary tone antagonized the bishop and the local pronuclear *senyors*, traditional stalwarts of the church, who repudiated him by attending Sunday mass in neighboring villages.

In the mid-1970s, Ascó was already a village profoundly split in two, the sides known to one another as the *pronuclears* and the *antinuclears*. The genealogical and sociological continuities between the civil war and the division over *la nuclear*—as the plants are popularly referred to—made the situation even more complex and bitter. The big landowners, whose land had been expropriated during the war and whose authority over the village had been strengthened during Franco's regime, were overwhelmingly on the pronuclear side, and so was the fascist mayor. On the other side, most of the leading figures of the *antinuclears* were members of households that had sided with the defeated during the war.[54] The internal social divisions over the nuclear issue were represented by the via crucis—a Catholic ritual recreating Christ's way to the cross—that Redorat organized in the summer of 1975 to commemorate the anniversary of the dropping of the nuclear bomb on Hiroshima. A look at this action will allow us to get a sense of the concerns, the ideology, and the increasingly hyperbolic rhetoric that pervaded the antinuclear struggle in Ascó. The event, which the pronuclear municipal authorities were not permitted to attend, started when a crown of laurel bearing the words "Ascó-Hiroshima" was thrown into the river. Then the priest gave a brief speech:

> The masters [*amos*] of nuclear energy think in their own interest and all too easily forget about the interests of the people and the rights of all human beings, the first of which is the right to life. The presence of nuclear power

plants in Southern Catalonia will eliminate, due to contamination, our agrarian wealth.... Vandellòs, Ametlla, Ascó, a nuclear chain reaction is starting to generate an anti-nuclear action to break the chains of oppression, of the exploitation of some by others. There are too many crosses in Hiroshima and Nagasaki. This is why a whole region [i.e., Southern Catalonia] is rising to defend its own survival in an act full of symbolism.[55]

After this introductory speech, the priest punctuated each of the fourteen ritual stations of Christ's journey to the cross with an analogy to the situation in Southern Catalonia: the suffering of Christ was likened to the degradation of the peasant, Roman imperialism to the capitalism of the rulers, and so forth. Perhaps the most revealing detail of the speech was Redorat's use of the word *amo* (master, landlord) to refer to those in charge of nuclear energy. The term pointed to the heart of both the intra- and inter-household power relationships structuring daily lives in agrarian communities of Southern Catalonia. As with the parallel emergence of peasant activism, it would be a mistake to assume that the priest was using traditional language that was incompatible with the new relations of power and production brought about by the nuclear power plants. On the contrary, his use of traditional language allowed him to uncover the connections between nuclear power and local class dynamics and thus build an encompassing discourse linking the antinuclear mobilization to a broader struggle against all forms of domination. The via crucis was an overarching denunciation of injustice and power abuses, a call for the creation of a new social order launched in a particular political climate—the waning years of the dictatorship—in which any change seemed possible.

By the time Redorat was pronouncing these words, most of Southern Catalonia seemed to be in a state of war against authority. Three key events that occurred in 1974 helped propagate insubordination and the dream of a new social order throughout the area. The first one was Hidruña's plan to revive the project of the Xerta dam, to which the electricity company now added a second one in the town of Garcia. In parallel, the government announced a related plan to transfer water from the river to the urban centers of Barcelona and Valencia, the petrochemical complex of Tarragona, and the steel-production center of Sagunt.[56] The latter two sites were industrial poles created in the early 1970s, star projects of the government's *desarrollismo*. These hydraulic proposals caused uproar throughout the lower course of the river. The antinuclear and anti-dam movements quickly developed a common understanding of their situation: "[Southern Catalonia] is equidistant from the advanced manufacturing centers of Barcelona and Valencia and from the heavy-industry subcenters of Tarragona and Sagunt. This central geographical position, together with the existence in this area of an increasingly coveted economic resource—water—are the reasons behind the reconversion of this part of the country into a supplier of energy and water to the industrial centers located to the north and to the south: nuclear power plants, water-transfer plans and hydroelectrical facilities all respond to this scheme ... [which] is part of a totalitarian capitalist industrial model."[57]

The second event, already examined in the previous chapter, was the foundation of the radical peasant union Unió de Pagesos (UP). Right from the beginning, UP took a strong position against the hydraulic and energy projects affecting Southern Catalonia, sharing with the antinuclear movement an analogous diagnosis of the situation of the rural world and the dispossession of its inhabitants. The local chapter of Ascó was one of the antinuclear bastions in the village, while in neighboring villages such as Fatarella and Pobla, UP organized blockades to impede the transport of materials to the nuclear construction site.[58]

The third key event that took place in 1974 was FECSA's announcement of a project to build two nuclear power plants in the fishing town of Ametlla, just fifteen kilometers south of Vandellòs.[59] Some elements of the antinuclear opposition in Ametlla were strongly reminiscent of the movement in Ascó: the struggle against a regime represented by the Francoist mayor, the involvement of a progressive priest, and the defense of the primary economic activities of the population. Indeed, the most visible leaders of the movement were Enric Rebull, president of the association of fishermen, and his young son, an ordained priest with a degree in sociology. But whereas in Ascó we can see from the very beginning internal social fractures that would only become more ingrained with time, in Ametlla we see a town largely united in rebellion.

When the representatives of FECSA visited the town in 1974 to buy the land where the power plant would be built, they were met by an angry crowd, and subsequent events resulted in the arrest of Rebull and a communist activist. The day after the arrest, the fishermen declared a strike and ran the sirens of their boats, which were anchored in the sea, all day long. A few days later, when another representative of the company went to the village to sign the purchase agreement, a group of women, members of an ad hoc housewives association created to respond to the nuclear threat, assaulted him and took off all his clothes. Two of these women were sent to jail for a few days, and a third suffered a nervous breakdown that led to her hospitalization. In 1975, the fishermen commissioned a multidisciplinary team from the University of Barcelona to conduct a study of the impact of nuclear plants in the village, and the team concluded that they would negatively impact fishing due to the warming of seawater used to cool the reactor.

The antinuclear tide in Southern Catalonia reached its high point in 1977, a year marked by important victories—some of them partial, some of them definitive—for the movement. In the summer of 1977, the *antinuclears* of Ascó, with Carranza and Redorat in front, forced the resignation of the mayor and his team, who were accused of operating in the interest of FECSA rather than the villagers. An interim board, led by Carranza and with a clear antinuclear majority, took control of municipal affairs until the 1979 elections. Furthermore, the plans to build nuclear plants in Ametlla and hydroelectric dams throughout the lower course of the Ebro, together with the water transfer schemes, had been removed from the new PEN that the first democratically elected government of Spain, led by a center-right coalition, was drafting. However, the final approval of this PEN would drastically narrow the horizon of the antinuclear struggle in Southern Catalonia and seal the fate of Ascó.

The Ideology of the Antinuclear Movement: Economic Growth, Peripherality, and *Autogestió*

The Spanish antinuclear movement was a reaction against the sociopolitical order that implemented the nuclear program and against the growth model and class interests that this program served. The two main criticisms that the Spanish antinuclear movement leveled against this model—that it was irrational and authoritarian—were in fact the bread and butter of the international antinuclear movement. We see these criticisms in the work of contemporary US antinuclear authors such as anthropologist Laura Nader.[60] On the one hand, the accusation of authoritarianism was central to the opposition to what Nader called "hard energy paths," representing "a continuation or elaboration of technologies such as nuclear, which would be centralized, authoritarian, controlling and not friendly to the democratic process."[61] Given that several decades of pro-market neoliberal policies and propaganda have accustomed us to a prostate inertia on the Left, it is important to emphasize that the antinuclear movement directed the charge of authoritarianism against both the state and monopolistic corporations, an effect of the intimate intertwining of state structures—including, significantly, the military—and private capital in the development of nuclear energy.

On the other hand, antinuclear intellectuals argued that nuclear power was put at the service of and derived its rationality from a fetishistic belief in an irrational notion of growth as an end in and of itself: "What are properly means—technology and the economy—have been elevated to the rank of ends."[62] And not only was the belief in autotelic growth irrational, so were its consequences: a wasteful economic system seeking infinite expansion and based on the depletion of nature, mankind immersed on "a drunken oil binge."[63]

Indeed, a major strength of the antinuclear movement, in the United States and elsewhere, was its capacity to link the two criticisms, establishing an internal connection between growth and authoritarianism. The Spanish antinuclear movement widely joined in this critique and was heavily influenced by its US counterpart, but went beyond it in building a common view that linked state domination with capitalist exploitation. The oligopolistic structure of the electricity sector, the concurrence of a democratic struggle against an authoritarian regime, and the semi-peripheral position of Spain within Europe all played a major role in propitiating the development of this understanding, but so did the deep trajectory of anarchist practice, the strong tradition of socialist libertarian thought, and the pervasiveness of the struggles for regional autonomy and local rule in the country.[64]

It was in uncovering what the movement labeled "internal colonialism"—a term with Gramscian and Lefebvrian echoes—that the Spanish antinuclear movement was most original and more able to build connections with the myriad local struggles spread throughout Spain.[65] In the dust jacket of what can be considered the foundational text of the Spanish antinuclear movement, the voluminous collective book *El Bajo Aragón expoliado*, edited by Mario Gaviria, we can read: "We observe a strategy of internal colonialism geared toward the extraction of natural resources (water, coal, hydroelectricity) and to the use of the territory as the basis of polluting activities (coal and nuclear power plants). This is how the state operates in the capitalist context: subordinating the interests of certain areas of the country, dependent or colonized, to other areas, the dominant ones."[66] Adding later: "The economic miracle of these [central] areas ... is based on spoliation, this is why the survival of some depends on the spoliation of others."[67]

Economic growth is therefore understood not as the product of capital's creative capacities—the result of economic dynamics such as labor productivity or the compulsion to compete—but rather the consequence of extra-economic mechanisms of dispossession. The vision of economic growth as a zero-sum game resonated powerfully with the experience of the peasantry that populated the rural territories targeted for nuclear development, as it can be appreciated in a song that a journalist from Ascó wrote in 1974: "*Amos* of the *camp* [fields, countryside], we have nothing: // no land, no water, no money. // Others dispose of our wealth. // For highways, nuclear plants, // dams and transfers // they keep stealing our land and homes."[68]

The antinuclear critique, however, was not limited to an indictment of the Spanish state of affairs; it also contained a positive, propositional aspect. This positive dimension may be summarized in the movement's demand for "rational" spatial planning policies and the vindication of autonomy. It is not hard to see behind the planning demands of the antinuclear movement a classic anarchist vision that can be traced back at least to Kropotkin, with his desire to eliminate the division between country and city.[69] The rationality of the spatial planning policies demanded by the antinuclear movement lies in the fact that they would be oriented to correct the asymmetry, and ultimately eliminate the dichotomy, between central and peripheral territories on which the existing economic model was built.

The creation of a new energy model was central to this vision. Instead of large power facilities located in the periphery to feed the needs of industrial poles, the antinuclear movement demanded small power units oriented to local demand. Instead of a centralized energy system deepening the rift between centers and peripheries, the antinuclear movement envisioned a distributed energy system based on locally available renewable resources. This energy model would be the basis for the development of a whole new set of dynamics: the balanced distribution of the population in the territory, the reproduction of peasant livelihoods, the use of locally available resources, the decrease of energy demands, and a new productive structure based on smaller manufacturing units. Ultimately, thus, the antinuclear movement envisioned a world in which autonomy would replace subordination and dependence—a vision with strong commonalities with the historical struggle of the Southern Catalan peasantry. The manifesto issued in 1978 by the Third Assembly of the Spanish Environmental Movement neatly encapsulates this series of ideas:

> We refuse to identify progress as quantitative growth and all that it entails, which in fact is simply the progress of the dominant class. We reject the capitalist mode of production and we consider unsatisfactory all forms of state socialism and, in general, any socioeconomic formula based on productivism and the accumulation of power. We declare ourselves in favor of free energies and against civil and military nuclearization, for it not only involves a clear threat for life, but also a militarized, monopolistic, ultrahierarchical model of society that is incompatible with *autogestión*. Instead, we favor the autonomy of communities, the plurality of life forms and the *autogestión* of labor collectives.[70]

The notion of *autogestión*, or *autogestió* in Catalan (roughly translatable as "self-management"), is key to understanding the program of the antinuclear movement and how it was able to exert influence in rural territories such as Southern Catalonia, for it helps us specify the meaning of the search for autonomy that I discussed in chapter 1.[71] As Henri Lefebvre argues, *autogestió* heralds an ongoing, limitless project of radical democracy that can potentially affect all aspects of life: "Each time a social group (generally the productive workers) refuses to accept passively its conditions of existence, of life, or of survival, each time such a group forces itself not only to understand but to master its own conditions of existence, *autogestion* is occurring.... This definition also includes all aspects of social life; it implies the strengthening of all associative ties."[72]

Autogestió points, therefore, to a program of radical democracy, a struggle against all ties of dependence, characteristic of the variegated forms of popular mobilization that arose in the exceptional historical conjuncture of the *Transición* and the opening of political possibilities that it entailed. The *Transición* would end up suffocating these political possibilities and marginalizing the antinuclear program. Yet as we will see in subsequent chapters, the antinuclear agenda in Southern Catalonia—the rejection of economic growth as a goal, the quest for autonomy, the critique of "irrational" planning, the demand of territorial equilibrium, the value given to primary activities—stands as a reservoir of political ideas (an uncompleted past) and an unfulfilled political program (an impeded future). Indeed, the demand of dignity will be, to a large extent, a vindication of *autogestió.*

Nuclear Consensus

With these considerations of the simultaneously political and environmental goals of the antinuclear movement, let me return to the PEN proposal, and explain how, ultimately, the nuclear program enabled a top-down management of the political transition. The term *Transición* in Spain encapsulates at least three meanings that are hard to reconcile with one another. The first meaning expresses openness, highlighting a liminal period full of uncertainty and open to political possibilities. The second meaning expresses a sense of achievement, the idea that the shift from dictatorship to democracy was a gradual, bloodless, exemplary process in which all Spaniards were able to agree and move forward to do the right thing— that is, instead of killing each other. It is this second meaning that saturates the hegemonic version of the *Transición*, its *myth*, endlessly repeated by all governments and mainstream media, which depicts the period as a massive operation of consensus building managed by selfless elites, a "series of tectonic adjustments to society made possible, not by selfish class interests ... but by intelligent technocrats and the modestly gentle guiding of an enlightened monarch."[73]

Needless to say, such a teleological narrative relegates to the margins not only the openness expressed by the first meaning but also the class struggle, conflict, and violence that pervaded the whole process, especially in the first few years (1973–1977). Indeed, only after 1977—coinciding with the legalization of political parties and the first democratic elections, which were followed in 1978 by the passing of a new constitution—did a new consensus emerge, trumping a good portion of the radical democratic aspirations that had characterized the first years of the *Transición.*

The third meaning expresses a sense of frustration and disappointment at the fact that the process was a "false social change" that did not result in a radical rupture with the Francoist regime. In the late 1970s the process was already being criticized for not being a transition so much as a *transaction*, a deal between the politicians of the old regime and the representatives of the new political parties. Made under the auspices of the economic elites and the Francoist bureaucracy, this transaction presupposed three conditions: acceptance of the heir designated by Franco as head of state, amnesia about the past (specifically, about the crimes of the war and the dictatorship), and continuation of the economic status quo.[74]

The myth of the *Transición* thus conceals the process—the transaction—through which political openness was progressively trumped, moving from the first to the second meaning of democracy. Democracy, Timothy Mitchell writes, "can refer to ways of making effective claims for a more just and egalitarian common world. Or it can refer to a mode of governing populations that employs popular consent as a means of limiting claims for greater equality and justice by dividing up the common world."[75] This "dividing up" consists of separating those issues that are considered *political*—that is to say, open for negotiation and popular decision—and those that are deemed *technical*, that is to say, nonnegotiable, the subject of expert management and administered control rather than democratic debate.[76]

The political process leading to the approval of the first energy plan of the democracy (PEN-78) offers a clear example of the creation of a nonpolitical sphere. The PEN-78 looked eerily similar to the one approved in 1975.[77] Although some nuclear projects had been dropped, the new PEN reiterated that the nuclear option was the only way to achieve an increase in electricity production capable of sustaining the rate of economic growth. The parliamentary consensus built around the PEN-78—what I call the nuclear consensus—participated in the constitution of a new exclusionary political order and the maintenance of the economic status quo.

The PEN's arrival in Parliament was preceded by an intense media campaign designed to counteract antinuclear sentiment, which was at its historical peak.[78] The Atomic Forum—the nuclear lobby supported by the electrical oligopoly—hired the director of Spanish public television and campaign manager of UCD (Unión del Centro Democrático), the center-right ruling party, to run an image campaign that focused on two ideas: nuclear energy was a technically neutral option that would bring jobs, progress, and prosperity, and the nuclear opposition was radical, violent, and aimed to destroy the Christian values on which Spanish society was built. Mainstream media of the Left displayed a more moderate rhetoric but a similarly unequivocal position, as evidenced in the series of editorials that the newspaper *El País* ran in favor of nuclear energy while censoring antinuclear columnists such as Mario Gaviria.

At the political party level, the nuclear option garnered a similar amount of crossideological support.[79] The process was not too different from the one that Marx described in *The Eighteenth Brumaire of Louis Bonaparte*, in which an unofficial "Party of Order" crossed ideological domains and took the reins of the revolution.[80] The difference is that in Spain there emerged a "Party of Growth"—or perhaps it would be more accurate to say that the Party of Order found in the consecration of growth and in a shared *epistemology of growth* its common denominators.

Indeed, Left and Right defended the atom with similar arguments: "Without electricity of nuclear origin we would be condemned to underdevelopment," stated the conservative minister in Parliament during the preliminary discussions of the PEN-78.

Santiago Carrillo, the historical leader of the PCE (Communist Party of Spain), declared to the press, "A country that refuses nuclear energy is a country that refuses progress."[81] Left and Right employed the same mental framework and saw growth—economic development, increased energy requirements, technological advances—as an inescapable requirement for the achievement of their societal projects. Thus, although Left and Right had radically different class projects, they both believed that growth was indispensable to achieving them and that only nuclear energy could guarantee such growth.

This level of political consensus is all the more surprising considering the high level of popular contestation to nuclear energy, which showed how the new parliamentary system was increasingly sealed off from outside political struggles. Only the Socialist Party (PSOE) offered a less homogenous vision and incorporated within the social dissensus around nuclear energy, materialized in the tension between a pronuclear leadership and a social base filled with antinuclear militants. This tension was clearly palpable in the Jornadas de Política Energética organized by the PSOE in 1977. The document that came out of this symposium articulated the arguments fundamental to the antinuclear movement.[82] In the conclusions, it called for a moratorium that would allow for the nuclear issue to be decided by a broad national democratic debate open to the citizenry.

Once the PEN reached Parliament, however, it soon became clear that all political parties, including the PSOE, shared a very peculiar understanding of what a democratic debate should be. The socialist speaker, Javier Solana—who years later would climb to the highest echelons of the EU—was clear: the PSOE was in favor of the nuclear program but wanted it to be democratically managed and supervised. In practice, this perspective involved, first, sanctioning those nuclear plants that, like the ones in Ascó, were already under construction; and, second, calling for the creation of a parliamentary commission that, with the aid of "experts," would discuss the future of the nuclear program and the fate of those nuclear power plants that had little more than an administrative authorization (such as Ametlla's).

This positioning was paradigmatic of the nuclear consensus that was being developed by mainstream political parties, a consensus that showed how the new institutional framework created by the *Transición* pushed away the demands of the antinuclear movement, thus condemning it to the margins of the political game. The debate would be limited to the terms of the nuclear program, rather than focusing on its existence, for it was considered an accomplished fact and a necessity for growth: any voice of dissent was marginalized as inimical to the progress of the country. In addition, the debate would not be open to the citizenship or take place in the streets, but would stay within Parliament, where only experts and democratically elected political representatives would be involved. "Democratic debate" came to be thus defined in the *Transición*. Energy debates were thus delinked from class struggle and radical democratic demands, therefore shielding the bundle of interests structured around nuclear energy from these demands.

Until now I have underscored the broad parliamentary agreement around nuclear energy. But the discussion around the PEN-78 also revealed a point of disagreement between Left and Right. This sticking point concerned the role of the state in the energy sector. Both the socialist and the communist parties argued for stronger intervention by the state in energy matters, unsuccessfully demanding modification of the PEN-78 to include the creation of a national oil corporation, the nationalization of the electricity grid, and a much stronger state role in the production of electricity—50 percent according to the PSOE. Certainly, the vision of the energy system put forward by the parliamentary Left did little to decentralize it and was very far from the anti-statist ideology of the antinuclear movement and its demand for *autogestió*. But the members of the electrical oligopoly were also very concerned with these plans, fearing that they would be expropriated if the Left came to power.[83]

A minority saw the confrontation between the parliamentary Left and the oligopoly in a different light. In fact, a series of economists had already warned in the late 1970s that investment in the nuclear program would lead to the financial ruin of the electrical corporations and make the intervention of the state inevitable.[84] Even some electricity tycoons, such as Hidruña's president, went against their peers by arguing that the financial obligations imposed by nuclear energy were too high to be assumed by private capital and therefore required the support of the state.[85] Many in the antinuclear movement considered that the financial ruin the nuclear program would provoke was part of a deliberate strategy on the part of the members of the electrical oligopoly, who would ultimately end up asking the state for help in order to socialize their losses. Economist Salvador Martín Arancibia powerfully expressed such an idea: "The political opposition is acting like a puppet in the strategy of the big industrial and financial groups, which is leading the [electrical] system toward its nationalization. At best [the opposition] may make it happen sooner than later, or dress it with demagogy and present it as a victory of popular interests."[86] These words were premonitory.

Sabotage

In the 1980s, most countries in Western Europe and North America abandoned their nuclear projects. Scholars are divided on the main causes of the failure of the nuclear transition.[87] Whereas some emphasize the antinuclear movement and an increasing social awareness of the risks of nuclear energy, others locate the main cause of its abandonment in the cumbersome financial obligations of the nuclear industry, which made it unattractive to private capital, especially in the context of lower oil prices. Despite their differences in emphasis, authors agree that both sets of causes were intertwined and reinforced each other. Thus, for instance, increased public awareness of risks increased the costs of nuclear development by forcing the industry to adopt more sophisticated security systems.

These two sets of causes interwove in Ascó in especially complex ways, resulting in a resounding financial failure that ultimately would lead to FECSA's bankruptcy. Behind the costs that drowned FECSA, we find a series of actors who *sabotaged* the construction of Ascó's plants through varied, largely unco-ordinated, and often-contradictory strategies and practices. While these actors had different goals and rationalized their actions in diverse ways, they all forced the company to divert increasing amounts of time and capital that would ulti-mately prove critical for its economic viability. My use of the term *sabotage* draws on Timothy Mitchell's *Carbon Democracy*, where he describes it as "any maneu-ver of slowing-down, inefficiency, bungling, obstruction" that is, the "conscien-tious withdrawal of efficiency."[88] Although the concept was initially created by the labor movement, Mitchell applies it to a broader set of actions and actors (including governments and corporations). The key, for Mitchell, is the capac-ity of these actors to interrupt the normal functioning of a process strategically located within the productive system.

Three main actors participated in the sabotage of the nuclear plants in Ascó: the workers, the local antinuclear opposition, and the electricity companies or, more precisely, the bundle of business interests that operated through them. I will start with the latter. The main saboteurs of the Spanish nuclear projects were the capitalist interests that most avidly pushed them forward, that is, those who controlled the electricity corporations that had embarked on the nuclear adven-ture: the banks. Nuclear projects worked as prototypical large-scale projects that offered big business the opportunity to open a new frontier for capital accumu-lation. Such an expansion did, however, involve remarkable financial risks for the electricity companies, especially when rising oil prices had notably reduced their profitability and depreciated the Spanish currency.[89] In this junction of low profitability and increased financial exposure, the Spanish banks, traditionally the main stockholders in electricity utilities, began a massive divestment while also maintaining their majority on the executive boards of those companies. That position of financial irresponsibility allowed the economic elites in control of the electricity utilities to keep promoting the construction of nuclear plants while shielding their account balances from the losses accruing from the same con-struction. The main beneficiaries of the construction contracts were corporations in which the banks, and sometimes the electricity corporations themselves, had majority ownership.[90]

In these circumstances, the economic interests that controlled the electricity companies had little incentive to control the costs of construction—which, given the dispersed stockholding structure of these companies, would ultimately have an impact on the savings of the Spanish middle classes—and a great deal of interest in increasing the extraordinary volume of capital, commodities, and labor that were being mobilized and concentrated through the construction process. The controlling companies then profited by draining resources from the electricity companies: overinflating prices, deliberately dilapidating equipment, directing capital and labor to unrelated projects, and so forth. Such practices did not go unnoticed to some conscientious engineers who, in retaliation, filtered internal documents to the antinuclear movement and thanks to whom we now have conclusive proof of these "antieconomic" strategies.[91] Local populations were similarly offended by the wastefulness and absurdity of this behavior, as it can be appreciated in the following quote from Tolo, who worked in the construction of the plants for almost a decade:

> Near the end of my time working there we needed a lot of olive tree wood for a project. So I had a neighbor in the village who had just cut down a whole grove of olive trees, because I think that he wanted to plant cherry trees.... Well, it doesn't matter. My bosses went to see my neighbor's wood, thought that it was good, and gave the OK. A few weeks later my neighbor gave me the invoice for the wood so that I could take it to my bosses, but when they, the people from FECSA, saw it, they jumped: "No way, this is not enough, don't you see that they're going to laugh at us! Go and tell your friend that it has to be higher." So my neighbor doubled the price and then it all went well. This I'm telling you so that you can see that they didn't care, *they were throwing the money away* so that they could take some of it.

Workers participated in the sabotage perpetrated by management. This was most obvious in their engagement in clientelist relationships that allowed them to take part in the systematic diversion of money and resources that was built into the nuclear strategy. Francis's experience as a worker in the nuclear plant as, in his own words, "a sort of foreman in the carpentry section" exemplifies this practice. Francis—who worked in the construction for most of its duration and has several relatives currently working in the plant—brags that due to his right-wing militancy he had "friendships" with some of the people who held management positions in FECSA.[92] These friendships helped him establish strong relationships with both managers and engineers, and these relationships translated into economic benefits: "I was in charge of taking note of the extra hours worked for my whole shift in carpentry. I always charged sixty or eighty extra hours for myself, but I also helped and did favors for others. And every single day, once the other workers had returned from their break, I would go to a bar in Ascó and have a king's breakfast. Every single day."

Francis's abuses reflect the practices of management in miniature—practices in which he also took part. For instance, he explains that a large part of his time was spent in the construction of houses for engineers and managers: "One day an engineer came to me and said, 'I came to ask you a favor.' A favor! *I* had to give *him* 'a favor'? I mean, unheard of, but this is how it went. From Cambrils [on the Mediterranean] to Seu d'Urgell [on the Pyrenees], we would find these houses every few minutes!" Not only the labor, with generously inflated hours, but also the materials used to build these houses were taken from the construction of the nuclear plants. Expert clients like Francis were able to use their position to assume the role of patron toward his coworkers, actively working to control and receive increasing amounts of resources, further stretching the costs of construction. These clientelist practices clearly supported corporate sabotage, and in fact, as we will see shortly, FECSA promoted clientelist relationships in order to erode and defeat the local antinuclear movement.[93]

Francis' emphasis on how little he worked is typical. "Are you going to work or down to la nuclear?" was one of the recurrent jokes that locals coined to mock the lack of discipline and low workload at the construction site. This contrasted with the very high salaries being paid in the nuclear construction (see chapter 3). In the view of the locals, everything about the construction of the nuclear plants was excessive: *tot era un excés*. The careless mismanagement governing the construction of the nuclear plant also translated into forms of deliberate systematic negligence on the part of labor, referred to by locals as *sabotage* or *boycott*.

On some rare occasions, these practices are framed as forms of political resistance, as in the case of this worker from Gandesa: "I only worked there for three months, in the last years, because I didn't want to work in the nuclear plant, but I had been unemployed for a whole year... And yes, I toyed with the labels and put stones in the tubes and caused a couple of floods. There was a lot of *boycott*, of course, and I wasn't the only one. I mean, I had to work, but I also had to pay back." Most informants, however, describe these actions as having less purpose than the previous quote suggests, something closer to active carelessness and deliberate foot-dragging, as Blanca, then a young antinuclear activist, explains: "When the workers got back to the village after their shift and went to the café you would hear them talking about how they didn't do any work. They bragged about spending the whole morning having beers and then peeing and throwing the empty bottles in the frame [which was under construction] of the nuclear reactor. And if you hear this and think a little bit you start to panic, because you know, this could create air bubbles in the reinforced concrete frame, and we are talking about nuclear energy, I mean nu-cle-ar!"

These daily acts of deliberate negligence, with their slight Luddite overtones, appear as mirror images of the wasteful practices of management, distorted reflections of the logic that guided construction—its excesses and its potlatch-like destruction of value. Yet while these practices may be seen as what Max Gluckman called "rituals of rebellion" strengthening the existing nuclear order, they also contained an element of resistance. Thus, we should also see them as "weapons of the weak," to use James Scott's expression, rooted not only in a deep disdain for the corporation and the plant, but also in an economic calculation oriented toward maintaining the part of the excess flowing to the workers.[94] Indeed, these negligent work practices—which, as we will see in chapter 3, coexisted with organized forms of labor struggle—would prolong the construction process and the salaries associated with it, for everyone was well aware that once the plant started its normal operation, labor demands would be dramatically reduced.

Yet the agent that did most to delay the construction process was Ascó's antinuclear municipal government. The control of the municipal government gave the *antinuclears* a strong capacity to sabotage the nuclear plants, for they could not operate without municipal permits. The control of the municipality and the electoral process thus became a major problem for nuclear interests. As we may remember, in 1977 Ascó's antinuclear faction was able to force the resignation of the mayor and gain control of the interim board that replaced him. The antinuclears were able to use two issues to force the resignation: one, the presumed illegality of the municipal permit that allowed construction to begin, for it had been issued before the nuclear project had received governmental authorization; and two, the accusation that the mayor, in collusion with FECSA, was promising permanent jobs in the nuclear power plant in exchange for political allegiance, keeping "a list" of the names of those who would get the jobs. This second charge was based on numerous testimonies from local residents, including this one: "Some people close to the mayor came to my home and asked me to find a worker who wasn't a rebel and didn't speak coarsely, because once the plant enters in operation the first ones [to get a job] will be those who have not opposed it and who have not made any mess. They said that they already have a thousand letters of dismissal prepared. And I answered that if in order to work in the nuclear plant we have to lower our pants, they should not count on me."[95]

In 1979, Carranza became the mayor of Ascó after defeating by a narrow margin the right-wing pronuclear candidates in the first municipal elections held since the war. Using that position of power, the antinuclear leadership tried by all possible means to prevent the installation of the two nuclear power plants in the village. Leaders organized marches and demonstrations, commissioned independent studies on the risks of nuclear power, and, most important of all, stood firm in their decision not to issue the municipal authorization that the plants would need to operate.

The company reacted against the municipality in two ways. First, the owners of the power plants refused to pay taxes to the village, financially asphyxiating a municipality that had received an inflow of workers from all parts of Spain. The municipal government filed a lawsuit demanding the money that was owed—approximately 7 million euros—but it would not be favorably resolved until 1989. Second, the owners kept warning, directly or indirectly through the pronuclear leadership, that the economic bonanza would dry up if the *antinuclears* remained in power, while keeping alive the idea that only those with pronuclear credentials would be eligible for permanent jobs.

The evolution of the nuclear debate at the national level also weakened the position of the antinuclear municipal government. Until 1978, the antinuclear struggle was part and parcel of the broader democratic struggle of the country. However, the nuclear consensus built around the PEN-78 left the antinuclear municipality without any strong political support. Ascó was largely left alone, consumed by a terrible internal battle and becoming a symbol of resistance for those increasingly desperate sectors of the Left—anarchists, environmentalists, Catalan separatists—that did not have representation in the emergent parliamentary party system on which the new constitutional order of the Kingdom of Spain was to be built.

From Antinuclear Municipality to Nuclear Village

The difficulty of governing a divided village, the increasing isolation of the antinuclear struggle, and, most importantly, the pressure of the electricity companies' promises of jobs proved fatal for the antinuclear municipality. In the early 1980s, the pronuclear side suffered a series of defections. As Jordi Ferrús explains in his encyclopedic ethnography of the installation of the nuclear plants in Ascó, the issue of jobs came to dominate the battle in the village, progressively overriding all other internal contradictions and ideological standings:

> One single division structures the village: those who have permanent jobs versus those who do not, those that will have it versus those that will not ... those who depend on the nuclear versus those who don't and don't want to.... All the rest ... the insults, the legal complaints, the demonstrations, the anonymous letters, the public meetings ... just hide this basic structural reality. It may take the form of envies and vengeances, it may be expressed by defending the nuclear plant as "good" or attacking it as "bad," but the reason behind all this is not whether it is dangerous but whether it provides jobs or not.[96]

In the battle over jobs, the pronuclear side held all the aces, incarnated in "the list." Even though Carranza defeated the mayor in 1977 and the pronuclear candidates in 1979, he was never able to get rid of "the list," which was controlled by the pronuclear leadership. It does not really matter whether a tangible list actually existed or not, although most seem to believe that it did. The point of the matter is that "the list" powerfully represented the monopoly that the pronuclear leadership held over jobs—the most coveted resource and one that was to become increasingly scarce. From the local perspective, the best proof of the reality of "the list" was that the pronuclear leaders had good jobs in the nuclear plants and strong connections with FECSA, and few doubted that they would get permanent jobs once the plants began to operate. Their position as clients of *la nuclear* in turn allowed the pronuclear leadership to become patrons in Ascó.

Carranza suffered a decisive defeat in the elections of 1983 and was replaced by the pronuclear leader. Four out of the seven members of the new pronuclear majority had good jobs in the nuclear power plant. After the 1983 elections, the war between *antinuclears* and *pronuclears* was generally understood to be over. The latter had won, and Ascó would become a nuclear village. As a result, approximately forty antinuclear families, an extraordinary number in a village of less than two thousand inhabitants, abandoned Ascó throughout the mid-1980s in what they described as "voluntary exile"—words that clearly resonated in relation to the war that had taken place half a century earlier. Carranza's family was the first to leave. In exile, Carranza was an unsuccessful candidate for a series of Green parties and died in the mid-1990s far from the Ebro. As for Redorat the priest, the bishop was finally able to force his retirement.

Once in power, the new pronuclear government gave municipal authorization to the nuclear power plants. However, the entry in operation of Ascó I was beset by a concatenation of technical failures that many blamed on the carelessness and negligence—the sabotage—perpetrated by the electricity corporations and, by extension, the workers. Up until this day, the nuclear plants in Ascó suffer the highest number of incidents and mishaps of any nuclear plant in Spain.[97] Initially planned to enter into operation in 1977 and 1979, Ascó I and Ascó II would not start producing electricity until 1985 and 1986, respectively, compounding FECSA's financial problems.

The end of the construction process did however pose a problem to Ascó's pronuclear municipality. Their victory had been built on a clientelist scheme that could only operate if patrons delivered, maintaining a steady flow of resources. But with the start of operation of the plants, the number of permanent jobs that the nuclear plants could deliver to the residents of Ascó was fairly limited.[98] Indeed, each plant only needs a large, unskilled labor force (between five hundred and seven hundred workers) once every fifteen months, when the reactor needs to be refueled, an operation that takes a few weeks.

In response to this contradiction, the municipality would come to substitute for the nuclear plants as the source of jobs in the town. With the pronuclear victory, the nuclear plants initiated a generous program of sponsorships to all civil organizations in the village (football club, historical society, etc.) and, most importantly, resumed paying taxes to the village, thus allowing for a dramatic increase in the municipal budget, which is nowadays on average ten times higher than those of neighboring villages with similar populations. Paradoxically, the municipal budget would also benefit from the echoes of the antinuclear struggle. Thanks to the lawsuit filed during the antinuclear legislature, the nuclear plants were legally obligated in 1989 to pay the taxes owed in the interim period. Furthermore, in 1990 the Spanish government established the "nuclear canon," an annual fee paid to all local municipalities situated in the vicinity of nuclear plants (see chapter 3).

All this income led to an exorbitant expansion of the public sector in the village, mostly funneled through urbanization projects, which came to satisfy the population's thirst for jobs. Not being overtly antinuclear was an unspoken requirement to benefit from the jobs provided by the municipality, the new patron. Today, the town of around sixteen hundred inhabitants employs around one hundred permanent workers, in addition to those with temporary contracts or those who work for subsidiary municipal companies.

Moratorium

In 1982, the Left reached power through the landslide victory of the PSOE, a momentous event that put an end to the *Transición*. Once in power, the socialists announced the drafting of a new PEN, passed in 1983. The PEN-83 introduced a series of measures that were absolute novelties in Spanish history.[99] Without a doubt, the European Union, of which Spain would become a member in 1986, was a major influence on these new directions. For the first time ever, emphasis was put on energy efficiency and conservation, whereas the construction of new facilities for electricity generation was discouraged. The PEN-83 nationalized the electricity grid and created a national petroleum company while establishing a nuclear moratorium that adjourned sine die not only all future nuclear projects but also a handful of projects that were in the early stages of construction.[100] The passage of this moratorium was fundamentally motivated by the incapacity of electricity corporations to meet the financial requirements that the construction of nuclear plants imposed on them.

The nuclear moratorium was a great victory for the antinuclear movement: no nuclear plants have been built since the 1980s, and the number of nuclear plants that have entered in operation in Spain—ten, four of them in Southern Catalonia—pales in comparison to the thirty-five proposed in 1975, yet the country never "plunged into darkness." However, as the antinuclear movement had long feared, the nuclear moratorium must also be seen as a public bailout of the electricity sector. Thus, with the moratorium the government established a yearly payment to the electricity corporations to compensate them for their nuclear investment. This payment was not made by the state but by the Spanish consumers, who up until 2015 contributed to it through an extraordinary charge in their electricity bills. The state also created a public company to take care of nuclear waste management, also paid for through a special charge on consumers' electricity bills. The final result was the massive transfer of wealth from the state and the Spanish citizenry to the holders of economic-cum-political power.

In parallel to the passing of the PEN, the government restructured an electric sector ravaged by the "nuclear fever," and even the most recalcitrant pronuclear media outlets at this point admitted that nuclear power was the "cancer of the electricity companies."[101] To fix the situation, the state promoted a series of mergers and acquisitions, further concentrating the sector and forcing the electricity corporations to exchange assets between them. Given the corporations' lamentable situation, the main beneficiary of these exchanges of assets was the public company ENDESA, which in 1984 became co-owner of FECSA's nuclear plants. In 1987, FECSA was ultimately unable to repay the massive loans that it had taken out in US dollars to build the plants in Ascó and fell into insolvency. The CEO of the company was eventually forced to resign while the central bank removed it from the stock exchange, fearing financial collapse of the whole sector. The minister of industry admitted that FECSA's collapse would have a tremendous impact on "the mass of small investors," for the main Spanish banks had long since reduced their holdings in the company.[102] FECSA's progressive absorption by ENDESA culminated in 1996, just a few months before the government carried forward a new reform of the sector that would lead to the privatization of ENDESA. These events, however, are already part of a different story (see chapter 5).

As an energy transition, the nuclear program was a resounding failure. Nuclear energy did not become the main source of electricity generation—let alone of energy supply—in Spain. Instead, it almost bankrupted the electricity sector, and new nuclear plants are unlikely to be built in the country in the foreseeable future. However, this failed nuclear transition allowed the oligarchic interests vested in the electricity sector to successfully navigate the political transition that began with the death of the dictator. At a key political juncture, the nuclear program offered the economic elites of the country, who had achieved their status during and in alignment with Franco's regime, an opportunity for their reproduction within the new regime. The nuclear consensus that fueled the nuclear transition was fundamental to the creation of the *Transición* as a change without rupture, based on a transaction between old and new elites. Even the nuclear moratorium and the subsequent restructuring of the sector are crucial moments in this transaction. The oligopoly was kept in place, although a traditional member—FECSA, the great sacrifice—was eventually absorbed by a public company, ENDESA, which thus became the sector's leader. Ultimately, the nuclear program was central in reproducing a certain economic model and a certain way of making profits that would continue to shape the form of Spanish capitalism: profits based on rent-seeking mechanisms, large-scale projects, regulated sectors, and oligarchic interest groups sharing strong positions in the high spheres of finance and the state.

The Shine of *la Nuclear*

Everything in Ascó looks shiny and new. Ascó is the only locality in Southern Catalonia where the football stadium has natural turf (a very expensive and environmentally unsound luxury in this dry area). Everyone in Southern Catalonia knows that if you want to attend a concert by a famous pop or rock band, the only option, apart from the provincial capital, is to go to Ascó. It is the shine of *la nuclear*, an overwhelming presence that hides any traces of the past. Indeed, in Ascó the past seems not to exist: the old village has largely been abandoned; agriculture has become a marginal activity; and no visible signs point to the feud between *pronuclears* and *antinuclears*. It is as if the memories of the nuclear conflict and the antinuclear struggle had to be erased at all costs. Speechless gestures, trivial jokes, unusual verb tropes, and circumstantial oral pauses uncomfortably fill the space of a nuclear fear that is rarely voiced.[103]

But if the glare of Ascó is not fully capable of blinding its past, it is even less successful at concealing its troubling future. In the 2000s, the municipality built an industrial park next to the plant, but most of it is empty: it hosts only two subsidiaries of the plant and a car wash. People in Ascó fear the uncertainty of a postnuclear future. They often express this fear by comparing a future of economic development that has not arrived and an agricultural past that has been repudiated. In the early 1990s, the municipality undertook to extend the irrigation system, a historical demand of the villagers. However, the progressive abandonment of the irrigated lands has led to a dramatic increase in the cost of water, and the *comunitat de regants*—"community of irrigators," the organization of landowners that manages the water—is fighting against bankruptcy. Eduard, a thirty-year-old man who works in a car repair shop and a member of the community, puts it boldly: "There is no perspective in Ascó. They talk about attracting industry and all this, but the fact is that nothing comes here. I mean, the only possibility that I see is returning to agriculture, I mean, I don't see people very interested in it, but what else can we do? But the fact is that the money that people have made has been sent out—to buy houses on the beach or in Barcelona, to go to study in the city— but nothing and no one comes back."

Jeroni was one of the first to go to university; when he got his degree in history in the mid 1980s, he became one of only three villagers to possess a bachelor's degree. Now he lives on the Atlantic coast of Spain and visits the village at least once a year: "This is a dependent village [*poble*], a subdued people [*poble*]. And we are bitter, we are bitter because we know that we have no future, that there is nothing after *la nuclear*, and that that thing down there [i.e., the power plant] is going to end sooner than later. Here, land has been abandoned, and once you abandon it you don't go back to it."

In Ascó, the victory of the *pronuclears* was total. Yet in the penumbra of the nuclear plants a different reality took shape. It is to this other reality that we will now turn.

Notes

1. Discourse given in Parliament on October 26, 1978; quoted in Martín Arancibia (1979: 293).
2. Quoted in White (1943: 351).
3. Debeir et al. (1991: 171, emphasis added).
4. On *regeneracionismo* and Spain's water politics, see Swyngedouw (2015).
5. Núñez (1995).
6. Alayo (2011).
7. See Terradas (1979) for a discussion of this pattern.
8. Hughes (1983); on the distinction between concentration and centralization, see Marx (1976: 772–781).
9. On Montañés and his relationship with Pearson, see Roig Amat (1970). There is a succinct but fairly detailed biography of Pearson online; accessed October 3, 2017, http://www.biographi.ca/en/bio.php?id_nbr=7645.
10. Roig Amat (1970: 254).
11. Roig Amat (1970: 253).
12. Roig Amat (1970: 254).
13. Roig Amat (1970: 255).
14. On the history of the Barcelona Traction, see Capel (1994).
15. Roig Amat (1970: 276).
16. López Linaje (1979).
17. Vilar (1964–1968, Vol I: 368); see also Vaccaro and Beltran (2008).
18. Capel and Urteaga (1994: 53–74).
19. Núñez (1995: 62).
20. Martínez-Val (2001: 280–283).
21. On the electricity sector during the Francoist regime, see Gómez Mendoza, Sudrià, and Pueyo (2007).
22. Capel and Muro (1994).
23. Quoted in Swyngedouw (2007: 21; emphasis added).

24. For an overview of these hydraulic projects in the Ebro basin, see Pinilla (2008).

25. Boquera (2009: 39); see also Moncada (1995).

26. Garcia (1997: 133–141).

27. On this point, see Lefebvre's discussion of the relation between energy, the state, and the production of abstract space (Lefebvre 2009).

28. Serrano and Muñoz (1979: 129).

29. Muñoz et al. (1979).

30. Serrano and Muñoz (1979: 173).

31. Swyngedouw (2007: 18–20).

32. Martínez Gil (1997).

33. Sudrià (1987).

34. Sudrià (1987: 343).

35. Mitchell (2011: 29–31).

36. Fernández (1999: 115).

37. Martínez-Val (2001: 315–316). On nuclear energy and nationalism, see Hecht (1994).

38. On these financial conditions, see Debeir et al. (1991: 171–191).

39. Cuerdo Mir (1999: 162–164).

40. Naredo (2009); see Muro (1994) for the case of FECSA, which was the most extreme example.

41. Martínez-Val (2001: 324).

42. Ribeiro (1994), Naredo (2009), and Flyvbjerg (2005). Roy (1999) and Whitehead (2010) develop similar analyses applied to the Narmada Dam, in Northern India.

43. Ribeiro (1994: 157).

44. Fernández (1999: 44–80).

45. Fernández (1999: 120).

46. Muro (1994: 84).

47. On the complex relation between the so-called energy crisis of the 1970s and the growth of nuclear energy during that decade, see Caffentzis (1992), Mitchell (2011), and Huber (2014).

48. There are two fundamental bibliographical sources for the analysis of the nuclear history of Ascó. The first one is the monumental dissertation of anthropologist Jordi Ferrús (Ferrús 2004), born and raised in the village. The second is a series of passionate books published by the antinuclear journalist Xavier Garcia (1997, 2008, and 2013). My account is heavily indebted to their help and contributions.

49. Garcia (2013: 53).

50. Ferrús (2004: 228).

51. Quoted in Garcia (2008: 44).

52. Midnight Notes Collective (1992: 194).

53. Published in *Triunfo*, February 2, 1974.

54. In Ascó the internal division before, during, and after the war was singularly acute, see Bailey (1990).

55. A description of this *via crucis* is published as an appendix in Garcia (2008: 148–151).

56. Alonso (1978).

57. Quoted in Alonso (1978: 33–34).

58. Benelbas, Garcia and Tudela (1977: 113).

59. Rebull (1979: 111–183) offers the best summary of the mobilization in Ametlla.

60. Most of Nader's contributions on nuclear energy can be found in Nader (2010).

61. Nader (2010: 519); the concept was developed by Lovins (1977).
62. Nader (2010: 540).
63. Nader (2010: 525).
64. On the environmentalist dimension of Catalan anarchism, see Masjuan (2000). For a discussion of the intermingling of political and economic functions in statecraft, see Jessop (2008).
65. See, for instance, Gramsci (1957: 28) and Lefebvre (1978: 173–174).
66. Quoted in Fernández (1999: 73–74).
67. Gaviria (1977: 93).
68. Quoted in Garcia (2013: 59–60). On the peasant notion of zero-sum growth, see Trawick and Hornborg (2015).
69. Especially Kropotkin (1985); see also Morris (2014).
70. See Varillas and Da Cruz (1981: 53).
71. On the impossible translation of the term, see Brenner and Elder (2009).
72. Lefebvre (2009: 135).
73. Narotzky and Smith (2005: 178). On the *Transición* as myth, see Gallego (2008).
74. Naredo (2001: 163).
75. Mitchell (2011: 9).
76. On this distinction, see Firat (2014). See Franquesa (2013a: 31–67) for a wider discussion of the evacuation of politics from the "streets" during the *Transición*.
77. See Cuerdo Mir (1999: 165–167) for a concise comparison of the two PENs.
78. On this campaign, see Pillado (1979).
79. Unless stated otherwise, my analysis of the debate around the PEN-78 draws on Martín Arancibia (1979; see also Rebull 1979: 50–67). Only the Basque parties broke the nuclear consensus.
80. Marx (1963).
81. Quoted, respectively, in Garcia, Rexach and Vilanova (1979: 32) and in Fernández (1999: 126).
82. Reprinted as Asamblea de Cercedilla (1979).
83. Martínez-Val (2001: 321).
84. See Serrano and Muñoz (1979) and Mestre (1977).
85. Ferrús (2004: 818).
86. Martín Arancibia (1979: 301).
87. See, for instance, Podobnik (2006) and Melosi (2006).
88. Mitchell (2011: 22).
89. Serrano and Muñoz (1979).
90. Serrano and Muñoz (1979: 242–267) offer the most systematic presentation of this data.
91. Naredo (2009: 51, n9).
92. This is a point that recurrently emerged in my interviews. Some managers used their position to organize right-wing party militancy in the region, a circumstance that allowed some of them to jumpstart relevant political careers at the national level.
93. See Gellner and Waterbury (1977) for a classic, contemporary study of clientelism in the Mediterranean.
94. Gluckman (1963) and Scott (1985).
95. Cited in Ferrús (2004: 393).
96. Ferrús (2004: 565).

97. See, for instance, *La Vanguardia*, January 5, 2015: "Ascó, Almaraz y Vandellòs, líderes en incidentes nucleares en 2014." See Ferrús (2004: 820–981) for a detailed recollection of the nuclear incidents in Ascó up to 1996.

98. Initially, the nuclear plants employed around five hundred permanent workers, although a large percentage of these jobs were not available to the local population (Sorribes and Grau Folch 1989).

99. See Cuerdo Mir (1999: 167–173) for a summary of the contents of the PEN-83.

100. Although it falls outside the scope of this book, it should be noted that a series of attacks perpetrated by the Basque armed group ETA played a crucial role in the withdrawal of those projects.

101. See "Costes diferidos, el cáncer de las eléctricas," *La Vanguardia* February 14, 1987.

102. See "Trastornos del apagón eléctrico: las veleidades del mercado bursátil" (*La Vanguardia*, February 9, 1987) and "Los socios principales de FECSA ya no tienen casi acciones" (*El Periódico de Cataluña*, February 10, 1987).

103. On these mechanisms, see Zonabend (1989).

3 Nuclear Peasants

Nuclear power is ... a future technology whose time has passed.
—Hunter Lovins, *A World for Generations to Come*

IN THE TOWN of Ascó the establishment of the power plants produced what may be seen, on the surface, as a classic sequence of modernization: the rise of industry, the wholesale extension of wage relations, and the rapid abandonment of agriculture. Yet this transition relied on accepting an uncertain nuclear future as well as a whole new set of relations of subordination and hierarchy. Becoming a "modern," middle-class wage earner with a permanent job required adherence to the nuclear project and its future. In contrast, in the surrounding agrarian villages of Priorat and Terra Alta, situated in the penumbra of the nuclear plants, we witness the emergence of a different set of processes: erosion of relations of dependence, the creation of a popular antinuclear symbolic universe, modest but generalized agricultural investment, new forms of peasant organizing, a search for economic alternatives, and the vindication of a certain peasant identity. The strength of these processes in Fatarella, combined with its proximity to the nuclear plants (less than ten kilometers), converted the village into a symbol of the antinuclear movement.

Taken together, these responses constituted a loosely articulated yet consistent alternative project of social transformation, reasserting a regional development based on endogenous resources. This countermovement was built on a moral economy that idiosyncratically combined antinuclear themes—*auogestió*, the risks of nuclear power, its authoritarian and extractive character—with a longer trajectory of peasant struggle. It manifested as an effort to remain *pagès*, to reproduce a "peasant condition" that Jan Douwe van der Ploeg defines as "a struggle for autonomy and improved income within a context that imposes dependency and deprivation."[1] In practical terms, this moral economy was oriented, paraphrasing Eric Wolf, toward maintaining the nuclear economy "at arm's length."[2] In effect, its aim was not to disengage from the nuclear economy entirely, but rather to subordinate that economy to the needs of the local population and to direct its resources toward the reproduction of local livelihoods.

Democratic Awakening

Southern Catalans often describe the 1980s as a period of "opening." The term describes the institutional democratization that followed four decades of dictatorship, but also the exposure to new political currents and social trends, thus being the local expression of the broader process of politicization that took place during the *Transición*. I will describe this democratic opening as it took place in Fatarella, focusing on three specific processes: the election of municipal governments linked to Unió de Pagesos (UP), the creation of new spaces of sociability, and what I call the "rebellion of the youth."

As in many other Southern Catalan villages, in Fatarella the first democratic municipal elections (1979) were won by an electoral list assembled by the local chapter of UP—a popular front against the remnants of Francoism, the latter represented by a series of right-wing parties. This party would maintain the political hegemony in the village for more than three decades, winning all elections until 2007 (see chapter 6). Most of the members of the new municipal government were very young and connected to the nuclear resistance, either as peasant activists or union leaders.

During the early 1980s, this municipal government undertook several lines of action. One was to secure the population's access to basic services. "They made a nuclear power plant as if it were in the middle of nowhere, with all that high technology, and here we didn't even have water, so we tried to get that," Lluís, a member of the first democratic municipality, told me as he denounced the extractive character of the nuclear economy. The municipality fixed the water supply problem, dignifying the village and its inhabitants without—as they were proud to mention—asking for any favors or money from the nuclear companies. In doing so, the local government asserted its autonomy, rejecting dependence on or fraternity with the nuclear plants. The unwritten principle of not asking for or accepting favors from the nuclear plants stands to this day.

A second area of municipal action was to establish an antagonistic position toward *la nuclear*. Like most neighboring villages, Fatarella installed a meter to monitor nuclear radiation. But unlike its neighbors, where these devices were rarely accessible to the public, Fatarella installed the meter in a visible spot of its town hall. More tellingly, the municipality organized a series of "ecological fairs" that showcased local agricultural production alongside antinuclear propaganda and exhibits on solar panels and wind turbines. Relationships built through the antinuclear movement were key in these events' organization. For instance, the fairs featured talks by university professors involved in the movement, the urban youth attendees knew the area through participation in antinuclear events, and renewable energy technology was brought to the village by cooperatives that emerged out of the antinuclear movement, such as Tecnolia (see chapter 5). Thus the fairs not only bridged peasant activism and antinuclear struggle, but also aimed to capitalize on the struggle by connecting rural agrarian producers with urban, alternative consumers.

Finally, the municipality worked to democratize the village's institutional structures and social life. The best example was the construction of a *casal*, perhaps best described as a popular community center. The *casal* was built by the villagers through *jornals de vila* (unpaid work devoted to collective projects) and largely funded by household donations. The *casal* hosted a series of local associations (hunters' club, folk dance groups, etc.) and had a huge bar, as well as a big room to celebrate concerts, conferences, and theater plays. It began to host an annual "village assembly," wherein the members of the municipal government presented the budget and responded to citizens' questions. As bars are important centers of sociability in Southern Catalonia, the *casal*'s bar had a profound impact on the life of the village. Fidel, a local antinuclear activist, argued that the community bar disrupted social hierarchies: "In the other bars, no one mixed. Everyone knew which tables they could sit at and which ones they could not—all divided by classes. And if you were poor, you knew you were paying more for your drink than if you were powerful."

Each of these initiatives was part of UP's struggle for dignity, its project to "convert the countryside into city" by elevating rural dwellers into full citizens who could enjoy basic services such as running water. This struggle for dignity also projected inwardly, to the social life of the village, for the creation of spaces of democratic sociability was crucial in undermining the political importance of differentiated privilege. Pep, a construction worker in his sixties, told me, "Dignity is like you and I, here, just sitting down. No one is on top, or below. No one is better for having been born one way or the other. Just the opposite. A person that is honest and sociable with everybody else, a good person." The search for dignity involved a combination of self-determination and solidarity.[3]

Eliminating privilege—or, conversely, extending it to the whole social body—was the basis for a dignified condition.[4] Ignasi, a wine grower in his twenties, said of the *casal* in his village of Corbera, built at the same time as Fatarella's, "This *casal* here does not have box seats ... because my father, who was a socialist, and some other people opposed [those boxes]. Don't forget that all these *casals* ... were built with *jornals de vila* ... so everyone is seated at the same level, no special box for the people of the good *cases*, because when they say *good* they mean *rich*." This *symbolic rejection of inferiority* is a widespread feature in the Spanish countryside, where it has historically played a critical role in the struggle against exploitation and clientelist structures.[5]

The municipality was not the only factor behind the local awakening. During the 1980s, a majority of the village's youth underwent a political radicalization. This was most evident in the creation of a youth association in 1980, whose members were heavily involved in the antinuclear movement.[6] A street theater and animation group called *Mas de la Solfa*—roughly, "musical farmstead"—quickly emerged from the youth association and became a staple of antinuclear demonstrations and the germ of a municipal School of Music. The youth association created a library within the *casal* and organized conferences on issues including drugs, feminism, renewable energy, and environmentalism.

In 1985, some of the younger members founded a monthly magazine that they called *La Cabana*, a reference to a kind of traditional drystone dwelling that is especially widespread in the agrarian landscape of the village. In addition to duly tackling taboo topics (drugs, the civil war, and sexual relations), *La Cabana* covered local labor conflicts and reported on every meeting of the municipal government. Drawing contributions from a wide range of collaborators, the magazine paid paramount attention to peasant struggles and environmentalism, featuring regular columns on organic farming, renewable energy, nuclear risks, and agrarian policies and organizations. Although these youth initiatives aimed to transform social relations within the village, they also sought to remain attached to, and invigorate, a peasant identity.

Fidel, a member of the youth association, recalls, "We were the revolution within the revolution." Indeed, the democratic opening in Fatarella involved an erosion of traditional hierarchies, not only among *cases*, but also within them, between generations. Importantly, women were widely represented in the youth association, making clear that this "rebellion against the elders"—to borrow Epstein's idiom—had an important gender component.[7]

This transformation of public life to include women as full members was linked to new income opportunities. Cecília, a member of the youth association and editor of *La Cabana*, explains that the opening of a textile factory made an enormous difference in the life of young women: "It gave a lot of dynamism to the village. We could have a salary without having to leave the village or ask your parents." Still, Cecília describes her work as highly exploitative: "The labor method was tough and sexist. I still have nightmares about it. The foremen were all men, and we were paid by piecework. I was constantly fined [for insubordination]. We tried to create a union, and the whole youth association fought for that, but it didn't work. Some women were boot-lickers." As we will see, the construction of the nuclear plants offered men a more advantageous situation.

Taken together, this collection of initiatives—*casal*, improvement of public services, ecological fairs, and youth organizations—worked to overcome the isolation of rural life, becoming a revolt against second-class citizenship that nonetheless remained attached to a peasant identity. This idea of an "awakening," of taking control of one's life and one's future, is perfectly summarized in the editorial of the first issue of *La Cabana*:

> There was a small, small village that lived (and lives) between drought and nuclear plants, in a time of crisis, tensions, transitions, constitutions,

elections.... But in the midst of all that, people who wanted to make their daily life better, who wanted to access that which was only available to the powerful or the urban dweller. Access to culture. Not an imposed culture, but a search for culture that could make us freer, that would make us not need shepherds.[8]

In short, the awakening was the outcome of the will "to make daily life better," to assert dignity.

Nuclear Workers, Antinuclear Peasants

The nuclear plants played a decisive yet ambivalent role in this struggle to make life better. In Fatarella, antinuclear positions became hegemonic, epitomized in the actions of the municipality, even as a large proportion of men worked in the construction of the plants. Antinuclearism and nuclear work maintained an intimate yet paradoxical relationship that gave rise to contradictory outcomes.

Nuclear Workers

During their construction, the plants' labor demands were formidable. For a decade (1976–1985), the construction site continuously employed between four thousand and six thousand workers. Between half and two thirds of this labor force was drawn from Southern Catalonia, occupying 50 percent of the active population of Ascó, and between 15 and 30 percent in the neighboring villages.[9] The rest of the labor needs were covered by outside workers, a large portion of whom were migrant workers from southern Spain, living temporarily in barracks on the site.

Initially, the nuclear site had trouble recruiting local workers. It could only attract peasants with little to no land. In 1974, Rafael, a young member of a poor *casa*, was the first person in Fatarella to go work at the construction site: "[A]t the beginning, working down there had a bad reputation. All those who had land did not want to work at *la nuclear*, it was seen as low." Labor conditions were not very good: Rafael earned fourteen thousand pesetas a month and had to walk to the site and back every day. In 1975, the company established a free shuttle service from Fatarella, enabling it to attract a larger number of workers from the village. These included Andreu (Joaquim's son, see chapter 1), who made twenty-five thousand pesetas as a skilled welder.

Noninheriting sons of middle *cases*, such as Ferran, were also drawn to the nuclear site: "My brother is the eldest, so the land was for him. And I was okay with that. I didn't want to be a farmer. I wanted to be carpenter, [so] the construction of *la nuclear* was perfect for me." In the late 1970s, the situation changed drastically: salaries spiked and the construction site began to draw local workers from all social backgrounds. Maurici, the only child of a middle *casa* from Marçà, in Priorat, remembers the shift: "You started hearing that many people are going to the construction of *la nuclear*, even people with a lot of land, which was shocking. So, working as a bricklayer I was making 20,000 a month, [and] you heard that the nuclear construction workers made 30,000 a week and 1,000 pesetas for every extra hour. Are you kidding me? Are you drunk? Nope. That was a different story, a different world."

Nuclear construction played a crucial role in the erosion of traditional hierarchies linked to land ownership, for it provided good, similar salaries to peasants of all backgrounds. Work in the construction site was especially attractive to young men, who constituted the backbone of the local labor force. With few exceptions, even the radicalized male youth of Fatarella worked in the nuclear plants—while making no effort, as we will see, to conceal their antinuclear opinions. In this respect, nuclear salaries also contributed to the youth rebellion that I described earlier, offering the male youth economic independence from their families.

It should be emphasized that working in the nuclear plant construction rarely involved abandoning farming. Nuclear workers farmed during the evenings and weekends. From the local perspective, this was made possible by the lax labor conditions and lack of discipline governing the construction site. The construction work was seen as physically undemanding, as we may remember from the joke that we saw in chapter 2: "Are you going to work or down to *la nuclear*?" On the other hand, collective endeavors, such as the construction of the *casal*, clearly benefitted from nuclear jobs. Low labor pressure allowed nuclear workers to get involved in community work, while good salaries translated into monetary donations. Ferran called the construction of the *casal* "the product of the moment: there was a lot of political energy, but it was also that money was flowing around."

In short, nuclear salaries undermined class differences and relations of subordination between generations; they provided crucial resources for collective projects and the enlarging of family patrimony; and they connected rural workers with industrial peers, all the while allowing for ample room to display antagonistic attitudes.

Antinuclear Leadership

The nuclear plants also contributed to the region's "awakening" by fueling the politicization of the local population and providing fertile ground for the emergence of political leadership. I will describe this process through the cases of three people—Ferran, Josep, and Fidel—who achieved positions of leadership and institutional authority in Fatarella. The three men are all part of the same generation, born in the early to mid-1950s. Their engagement with the nuclear plants and their antinuclear stances were dissimilar: Ferran as a nuclear worker and union leader, Josep as a peasant activist and an environmentally minded farmer, and Fidel as an antinuclear activist and cultural agitator. And yet, *la nuclear* provided the common denominator allowing their respective political impetus to converge and coalesce.

To understand Ferran's involvement, it is first necessary to go back to the salary increases of the late 1970s. The company's saboteur disregard for its own finances and its will to gain ascendance in the region (see chapter 2), as well as its need to attract a reluctant local labor force, help to explain these increases, certainly. Nonetheless, the most decisive cause was the sharp growth of organized labor unrest in the construction site. This unrest must be put in a broader historical context. After Franco's death, labor struggle escalated all throughout Spain, giving rise to an unprecedented increase in industrial salaries and a spectacular redistribution of wealth between 1975 and 1980.[10]

One of the most obvious indicators of the extent and political character of this labor struggle was the dramatic increase in the number of strikes in the construction sector, which had historically been poorly unionized.[11] Large-scale projects became strategic points of struggle for labor unions, and Ascó's nuclear plants were among the main targets, hit by a series of massive strikes between 1977 and 1979.[12] Local antinuclear resistance, the democratic opening, and the emergence of peasant activism coincided with, and contributed to, an intensification of labor struggle.

Labor organizing at Ascó was led by outside workers, who possessed more experience in industrial labor relations and labor activism than the local labor force, as Ferran admits: "We had no idea what a labor union was, let alone a strike!" However, he would soon get involved in the organizing efforts, as would others, becoming key mediators between the unions and the local labor force. Several local union leaders ended up occupying relevant political positions in their villages, a circumstance that Ferran explains with a favorite mantra: "*La nuclear* was a school."

Ferran was the first president of the *casal*, a councilman during the 1980s, and two-term mayor of Fatarella in the 1990s. Once the nuclear construction ended, he opened a small workshop, working independently as a carpenter. Although he never again worked for a salary, he stresses: "Thirty years later, I keep thinking as a worker, and I learnt it there. I can't think as a businessman; I can't stand workers being disrespected. I can't feel superior to them. And if you ask me where this comes from, well, it all comes from Ascó. It affected me a lot: you had to defend yourself as a worker. Those were difficult times, like today after all, with lots of scalps, the company against you..." In short, for Ferran, being *antinuclear* meant fighting against exploitation as a nuclear worker.

Josep was the inheriting son of a middle *casa* and one of the few men of his generation who never worked in the nuclear plants' construction. In 1975, he became one of the first villagers to join UP. He was motivated to defend the agrarian sector and to participate in the political struggle against the dictatorship. Through his involvement with UP, Josep learned about the threat that *la nuclear* posed to the environment and local agriculture. Returning from military service in 1979, he became a founder of an antinuclear commission within UP and a councilman in Fatarella's first democratically elected municipal government.

He spent the early 1980s arguing against the nuclear plants, insisting they were "an aggression against our territory," yet he recalls always meeting the same frustrating reaction: "Everybody agreed with us, and most people joined the demonstrations and roadblocks that we organized, but at the same time everyone kept working in the construction. You know, I understood it; especially those who were poorer needed the money. I never argued against anyone for this. But I said to myself that I would be coherent with myself and wouldn't go there."

In 1983, once the *antinuclears* lost control of Ascó's municipal government, Josep decided to convert UP's antinuclear commission into an ecological one. The struggle had to be reoriented: it was not anymore about trying to stop the plants, but about using the ecological sensitivity that antinuclear activist peasants had developed to create new agricultural approaches that challenged an official agrarian discourse that was "all about chemicals and productivity."

After several failed business attempts—such as introducing a new breed of sheep and creating an organic production coop—Josep and two other antinuclear commission members started a small company selling organic products (mostly nuts and olives and their derivatives) in 1987. They built a small factory in Móra and a shop in Reus where they directly sold most of their production. "I guess I've become a businessman," he says. "People may say, 'Look at this red, now he is employing people!' And I guess they are right, but I just tried to live off the land. And in fact I do it: I keep working the land of my *casa* with my father and we bring the production here [to the factory], and we also pay the farmers who sell to us well."

The son of a modest *casa*, Fidel showed his ability as a student from a young age. In the early 1970s, a scholarship allowed him the rare opportunity to attend university in Barcelona, where he studied labor law. The university in those days was a hotbed of political ideas, and Fidel became involved with several left libertarian collectives. In 1977, he began working as a labor lawyer in an extraordinarily contentious historical context. In parallel, he became progressively embroiled with the antinuclear movement, spending more and more time in his native village. To the dismay of his widowed mother, he returned to Fatarella in 1979, where he opened a law practice: "I realized that my struggle was here," he recalls succinctly. Back in Southern Catalonia, he became a major cultural agitator, founding the youth association, joining *Mas de la Solfa*, and helping to organize the ecological fairs. Later, he helped to create several short-lived environmentalist organizations and a still-standing countywide center for cultural research in Terra Alta. In the 1990s, he served two terms as Fatarella's vice-mayor.

Like Josep, Fidel spent countless hours campaigning against the power plants and shares stories of how the villagers helped craft banners for antinuclear events after their shifts at the construction site. Yet Fidel did not feel frustrated. For him, this apparently paradoxical behavior revealed a sense of certainty, an unequivocal sign that *la nuclear* was being defeated in the ideological sphere. Antinuclear people kept working in *la nuclear*: "Yes, it happened, of course, but we never saw it as incompatible. Here in the village, we understand, we've always understood this. People needed money ... and we always respected that, but that should not fool us, because a different thing is what is right or wrong [*l'altra cosa és les coses tal com són*], and people knew that."

As these three men's stories show, the locals' antinuclear impulses had multiple sources. They combined peasant concerns with environmentalist ideas, and labor struggle with the fight for democracy. This idiosyncratic blend would give rise to a particular moral economy—in Fidel's terms, a shared understanding of "what is right or wrong."

Antinuclear Peasant Moral Economy

The concept of *moral economy* was coined by E. P. Thompson and further popularized by James Scott's work on peasant resistance and rebellion in Southeast Asia.[13] The term describes shared notions of economic injustice and exploitation among subaltern groups, and the authors used it to underline these groups' agency and reveal their ongoing patterns of collective action. Scott argued that the violation of traditional values and moral entitlements triggered peasant rebellions, and he called for a phenomenological theory of exploitation primarily focused not on its material, objective dimensions but on the subjective perceptions of the exploited, crucially shaped by their culture.[14]

Drawing on these insights, I am interested in understanding how the Southern Catalan peasantry mobilized a dynamic combination of norms, meanings, and practices—a particular moral economy—to metabolize the patterns of capital accumulation and the structural inequalities generated by the nuclear economy. I characterize this moral economy as an *antinuclear peasant moral economy*.

Nonetheless, my approach differs from Scott's on two key axes that have largely dominated the research on moral economies.[15] On the one hand, the notion has been applied, almost exclusively, to the systems of provisioning of marginal or subaltern groups emerging in the interstices of dominant economic systems. This has obscured the fact that, as Palomera and Vetta put it, "all economies are moral economies."[16] That is to say, all economies, also the dominant ones, are embedded in social relations and institutions, entangled in value systems and cultural norms. Emphasis on the interstices has thus eroded the relational dimension of the concept, delinking the analysis of subaltern moral economies from that of the dominant moral economies "within and against" which the former operate.[17] The moral economy that emerged in Southern Catalonia was a reaction to another moral economy, that of *la nuclear*, and this is what made it specifically antinuclear rather than merely alternative to the nuclear economy.

On the other hand, the notion has largely been confined to the symbolic, often reduced to the operation of norms and values understood as the triggers of social action. This has emptied the concept of class content, thus obscuring the material bases of domination and resistance.[18] The moral economy that emerged in Southern Catalonia certainly drew on norms and values, on a certain symbolic universe configuring a cultural definition of "peasantness." Yet this moral framework, which eventually became hegemonic, emerged out of a specific set of material relations and in connection to a political project aimed at the reproduction of those relations while curbing dependence. It meant to rein in the extractive character of the nuclear economy, placing the latter in the service of the long-term reproduction of local livelihoods. This is what made this moral economy distinctively *peasant*.

The Morality of the Nuclear Economy

Ascó's nuclear plants—the bundle of corporate and state interests articulated around them—tried to establish their hegemony in the region through a specific moral framework attempting to constitute a particular sense of reality. Its central premise consisted of persuading Southern Catalans that the nuclear economy would "save" the region and its inhabitants through a future of development and prosperity that would supersede an old, stagnant agricultural economy. Notice that accepting *la nuclear* as a savior simultaneously involved accepting the idea that the region had little value and could not develop endogenously—that it needed saving.

These claims underpinned the attempts of *la nuclear* to disguise its domineering position as a moral relation, at once veiling and validating its domination in a process whose logic is described by Pierre Bourdieu: "To be socially recognized, [domination] must get itself misrecognized.... [R]elations of domination... must be disguised and transfigured lest they destroy themselves by revealing their true nature; in a word, they must be *euphemized*."[19] Indeed, accepting the idea that *la nuclear* would liberate the region from underdevelopment amounted to accepting that its dominant position was morally good. The nuclear economy *should* guide the destiny of the region and its inhabitants.

Local antinuclear forces always struggled against *la nuclear*'s strategies of misrecognition whether by calling *la nuclear* "master," tracing a connection between *la nuclear*'s authority and traditional relationships of dependence and subordination; by systematically rejecting any form of financial aid from the nuclear companies; by supporting labor struggle; or by fighting against the idea that one has to become pronuclear just to get a job. In this respect, the widespread adoption of antinuclear positions among Southern Catalans signals a failure of the process of euphemization on which the hegemony of *la nuclear* largely relied.

Take, for instance, the case of Miquel, a founder of *La Cabana* and second son of a middle *casa*, who worked in the construction site between 1983 and 1985: "After the military service, I worked there earning 90,000 pesetas [a month], and I had absolutely no skill. And I was the king of the castle! I bought a tractor for my parents, and I still had money left to live like a king." Like most young, local nuclear workers, Miquel made no effort to conceal his antinuclear opinions. His boastful disdain toward *la nuclear* puzzled some of his coworkers who came from outside the region: "While working in the nuclear plant some people inside said to me, 'What are you doing here if you are an antinuclear?' 'What do you want me to do?' I told them. Should I go to pick hazelnuts for the village *cacique*?"

This might seem to suggest a preference for the "nuclear *cacique*" over the "village *cacique*," but the issue is more complex. In fact, by making his antinuclear position public and bragging about his low skills and how little he worked, Miquel was able to symbolically reverse the position of dependence toward *la nuclear* into one of trickery, turning the *cacique* into a fool. This symbolic rejection allowed nuclear workers to make his work in the nuclear construction compatible with an antagonistic position that fueled labor unrest and the antinuclear movement.

On the other hand, the rejection of the morality of la nuclear was not without material consequences. Miquel's reference to buying a tractor for his parents is, in this respect, highly revealing: working in the construction of the nuclear plants allowed him to have fun, be autonomous, get involved in strikes and start a magazine with strong antinuclear opinions—*and* it also allowed him to strengthen the peasant economy of his household, the ultimate act of trickery. It is almost impossible to speak to any Southern Catalan about his experience in the construction of the nuclear plant without hearing how the salaries were used to improve his family's farming operation. Andreu is especially explicit: "All the *pagesos* who went down there got a big benefit from it. We fixed our houses, we bought land, and we bought tractors. And we worked with specialized workers, people who came from elsewhere to work in the construction, and they would come to you by the end of the month and ask for money, because they had polished it all off all with gambling, prostitutes, and drinking. But the mentality of the peasant from this area was more like, 'Let's take this, because it won't happen ever again.'"

The morality of *la nuclear* aimed to posit Southern Catalans as former peasants who should be grateful at the opportunity of becoming modern, industrial (and docile) workers. Southern Catalans rejected that morality by thinking of themselves as long-term peasants who took temporary advantage of the nuclear economy as indocile workers.

Extractive Economy, Barren Future

La Cabana is an unsurpassed archive of the demands, preoccupations, and projects of the villagers of Fatarella during the 1980s. In its pages, one can observe that the difficult situation of the local agriculture—stagnant prices, rising costs, drought, the uncertainties around joining the EU—was paramount in the village. One of its editorials finishes by pondering how to fix the situation: "Struggling, of course.... It will be a long and frustrating struggle, but this is the only way to live off of our labor, to live off the land.... We peasants cannot permit ourselves to be marginalized as the peasantry has always been."[20]

The magazine contributed to this struggle by building a symbolic universe that reversed the morality of *la nuclear*, as evidenced in a series of illustrated front covers contrasting two images. One shows a peasant strike in bright colors beside war material in dark tones. Another contrasts the village skyline during the *Festa Major* (summer fiesta), dotted with fireworks, with the skyline of the nuclear plant, encircled by radiation. A third contrasts the figure of death, hovering around a nuclear reactor, with the life a new irrigation scheme would bring to the area. And a final image opposes the demand for "work with dignity" (*feina digna*) in the new county hospital with the degeneration associated with the jobs offered by the nuclear power plants. In reference to the latter, the cover reads, "We don't want to be cannon fodder."[21]

The nuclear power plants are thus associated with danger, dependency, a barren landscape and a dismal future, whereas peasant activism, labor demands, village life, and farming are posited as a source of joy, fertility, hope, and dignity. These contrasts condense the normative kernel of the antinuclear peasant moral economy: local agriculture is a source of fertility that can generate autonomous livelihoods, but it is threatened by a nuclear economy that is as a source of dependency and underdevelopment. Local land contained hope; the nuclear economy was barren.

The emphasis on barrenness is key, because it situates the nuclear economy as an unsustainable, extractive activity unable to secure the long-term reproduction of local lives and livelihoods.[22] Indeed, the nuclear plants established in Southern Catalonia possessed some of the key characteristics of extractive economies: an enclave industry, connected to large capital networks, attracting an inflow of male, largely captive labor power to a relatively remote area; redirecting profits and output—electricity in this case—to urban centers while concentrating health risks and environmental degradation in the producing region; and huge income differentials combined with a disregard for existing economic activities endangering the continuity of a way of life.[23]

This extractive character becomes even more salient when we compare the nuclear economy to the region's farming, with its important component of self-provision and self-direction of the labor process, its careful management of the territory and, above all, its quest for intergenerational continuity. These characteristics are encapsulated in the term *cura* (care).[24] The people of this region believe farming should be about taking care of the land and the landscape, giving rise to fields, landscapes, and livelihoods that are *arreglats* or *endreçats* (well arranged, tidy; see chapter 7). This notion supposes a nurturing, reciprocal relationship between land and labor: if you take care of the land, the land will reciprocate. The ordering of the objects of labor (fundamentally, land) is performed so that they provide good yields while preserving their value, without exhausting them or letting them go waste.

Cura also refers to the quality of labor, and is crucially linked to the proper way of working, locally referred to with the expression *treballar a gust* (working willingly, at ease; see chapter 7). For instance, if your labor obligations do not allow you to devote enough time to farming, or if low agricultural prices squeeze your farming operation, you will not be able to "work at ease" and you will not be able to "take care" of the land. *Cura* thus requires a certain level of self-management (*autogestió*) and self-sufficiency, but it also strengthens these by ensuring the reproduction of the reciprocal relationship between land and farmer.

In contrast to the logic of peasant farming, extractive economies are discontinuous, both spatially—taking the form of enclaves—and temporally—generating boom-and-bust cycles. The temporal discontinuity was especially troubling and palpable in the early 1980s: the construction of the nuclear plants would end, sooner rather than later.

The moral framework presented by the covers of *La Cabana* aimed at strengthening locals' belief in their own resources, their self-esteem, reformulating the desire to remain *pagès* into a social and political project. Ploeg calls this effort the *peasant principle*: "The peasant principle contains hope ... [It] is about facing and surmounting difficulties in order to construct the conditions that allow for agency.... It links the past, present and future, it attributes sense and significance ... and it embeds the many different activities and relationships within a meaningful whole.... It stresses the value of working with living nature, of being relatively independent, of craftsmanship and pride in what one has constructed. It also centers on confidence in one's own strengths and insights."[25]

The "embedding of different activities and relationships within a meaningful whole" is central to the reproductive strategies of Southern Catalan households, almost invariably engaged in pluriactivity. Finding the right balance between these different activities and demands is, again, a critical task that also requires *cura*. As in farming, *cura* does not involve a disengagement from the market, but rather indicates the need to establish a certain distance from it—to keep it at arm's length—in order to find the proper balance for reproduction. The nuclear economy posed, in this respect, a fundamental challenge. The antinuclear peasant moral economy may be seen as an effort to harness the nuclear economy, to subordinate it to the reproduction of local livelihoods. The moral injunctions around nuclear money, the idea that it should be handled with *cura*, were crucial to this endeavor.[26]

Barren Money, Fertile Livelihoods

In 1986, just a few months after the Chernobyl accident, *Mas de la Solfa*, the street theater company at the vanguard of Fatarella's youth movement, put together a new show: *Askobyl*. The show opens with two peasants bucolically farming their land when they receive a visit from Uncle Sam, who starts throwing money at them. The peasants take the money, put on industrial clothes, and start building a nuclear plant, inaugurated with the presence of Church, Army, Government, and Capital. Then a nuclear accident occurs and destruction ensues: deformed bodies, aborting women, and death invade the scene.

Askobyl stands as a rather spectacular representation of the key themes of the antinuclear peasant moral economy, contrasting a peasant past, associated with life and fertility, with a nuclear future of death, barrenness, and authoritarianism. Money articulates the passage between these two pseudohistorical stages, but there is a catch: the money in the play is a hoax, green bills with a joker on one side and a deformed face on the reverse. The peasants-cum-nuclear workers were tricked. Thus nuclear money is false money, barren money that annihilates the social peasant order, along with its moral values and the value of its agrarian economy.

Local memories of the arrival of nuclear power are saturated with money, often described as excessive, impossible to digest. These comments tend to focus on morally reproachable consumption habits. Analogies with Hollywood's Wild West are a common trope, as in the following quote from Fidel: "Ascó was like a Wild West movie. Have you watched westerns? A lawless town, with gambling, prostitution, I think there were two pole dance places ... and then the famous gambling sessions on payday. It was mostly workers from elsewhere, not from the region, but some people from here also joined in." The iterative reference to the American frontier, crowded with uprooted characters, is telling, reminiscent of the commodity rushes and the boom-and-bust towns described by the literature on extractive economies. The pervasive insistence on gambling, drinking, and prostitution, too, is analogous to the stories that gloss the demise of big *cases* in the 1950s (see chapter 1).

Nuclear money materially encapsulates the extractive character of the nuclear economy. The wasteful consumption practices stemming from it stand as the exact opposite of *cura*, oriented to the preservation of local value and the long-term reproduction of livelihoods. Yet despite the emphasis that local memories place on excess money and binge consumption, most Southern Catalan nuclear workers siphoned nuclear money into their farming operations. This process of reinvesting nuclear money in agriculture reached its zenith in the mid-1980s, when the generous severance payments that accompanied the conclusion of the construction process even provoked a mild inflation of land prices.

Askobyl and the critical comments of my interviewees should thus be seen as attempts to shape and reinforce behaviors shielding the local economy from the dangers of the nuclear economy. Repeating these dangers is, in fact, recurrent practice in the peasant principle.[27]

In their classic essay on the morality of money, Bloch and Parry argue that money is, in and of itself, morally undetermined.[28] Yet the cross-cultural tendency is to understand money as morally opprobrious whenever it interferes with the reproduction of the household and the broader sociocultural order. Fables about nuclear money served that function, yet what was being debated was not the morality of nuclear money, but the way it was used. Was that money embedded in the long-term cycle of the reproduction of local livelihoods or was it *malgastat* (wasted, thrown away, literally, misspent)? Wasted money is that which does not preserve value: it does not contribute to reproduction and to the maintenance of the familial patrimony and the value of the local landscape.

For Bloch and Parry, most societies demonstrate an effort to subordinate the short-term cycle into the long-term restorative cycles on which the reproduction of society rests. This effort is materialized in a series of operations of conversion, by which the goods that derive from the short-term cycle are converted into durable resources for the reproduction of society. Elizabeth Ferry describes this mechanism in the Mexican mining cooperative community of Guanajuato, where silver miners reconvert the money generated by selling silver on the market into what Ferry calls "substances of place-making" (most notably children, respectable houses, and public architecture in the vernacular style). To do otherwise, "diverting silver money from its path back into cooperative patrimony—stealing or wasting it—is considered shameful for then it ... 'goes off into the air.'"[29]

In this respect, the moral framework I examine was especially concerned with the nature of these operations of conversion. The central danger of nuclear money was not that it would be "gambled" or "drunk," but that it would be spent on substances that destroyed the accumulated value of the village. Miquel expresses this idea in relation to the village's built environment: "Let me tell you: *la nuclear* is responsible for the destruction of the architectural heritage of this village. Yes. When the money started to arrive, people started to renew their houses and they committed atrocities. They did atrocities because they did not want stone, it reminded them of misery and of the cold of the *mas*; it was associated with dearth and with working for someone else.... Fatarella is still a pretty village, but it would be prettier were it not for *la nuclear*."

The goal was for nuclear money to be converted into "substances of place-making" that strengthened what Ploeg calls "a self-governed resource base," the material basis of the small producer's autonomy.[30] Maurici, the wine grower from Priorat, provides a particularly successful example of conversion. In the early 1980s, he combined the cultivation of his *casa*'s land with temporary wage labor in construction and agriculture. Although still in his early twenties, he was married to Josefina, and the couple lived in Maurici's *casa*. The *amo* of the *casa* was Maurici's grandfather, with whom he never got along. Political differences separated them: Maurici joined UP at its inception and would, in time, become a radical peasant leader with strong sympathies for Catalan separatism; his grandfather had fought in the Civil War alongside Franco's army and maintained strong right-wing views throughout his life. Maurici experienced their relationship as oppressive, a permanent curtailment of his autonomy: "I always wanted to do new things, but I felt tied down by my grandfather. I wanted to try new things, like plant new crops or even go organic, but he wouldn't allow me to." Maurici wanted to establish his own household, and for this he needed to buy a house:

> My partner [*companya*] and I wanted to buy this house [the house we were in and where they live] and we were thinking: "We'll have to go to the bank, because our families don't have the money, and they already paid for the wedding banquet." In that time you worked, had a girlfriend, and were ready to get married, but you gave your salary to your parents. And they bought your furniture, which is the furniture that they liked; they paid for the banquet and chose whatever they thought was fashionable at that time.... But we had different aspirations. So instead of going to the bank, I said, "Josefina, I am going to *la nuclear*." I worked there for four years.

Working in the construction of the nuclear plants liberated Maurici from dependence on his grandfather, without having to trade subordination to his grandfather for indebtedness to a bank. It gave Maurici autonomy, measured in distance both from the market and from his grandfather's authority. The house was an instrument to strengthen his autonomy, in a circularity that stems from the peasant effort to find a balance between market and reproduction through myriad processes of conversion between what we could call different "spheres of exchange."[31] Indeed, Maurici's house provided crucial added income by doubling as an agro-tourist residence—the first in the region—managed by Josefina, who insists that "agro-tourism is not like rural tourism: anyone can do rural tourism, but to do agro-tourism you need to be a *pagès*." By converting nuclear money into a house, Maurici and Josefina denied its all-out interchangeability and "singularized" it—to use Kopytoff's expression—by tying free-floating resources into the household and widening the self-governed resource base that is the ultimate guarantee of the peasant household's reproduction.[32]

Maurici and Josefina put nuclear money at the service of a reproductive logic alien to the nuclear project, converting nuclear money into the basis of their livelihood, while also strengthening the value of the village by engaging in the development of a new economic sector. The salary from the nuclear plant was in fact the beginning of a successful farming trajectory. If their agro-tourism business gave them important, complementary revenue, Maurici and Josefina saw the price of their produce increase exponentially at the turn of the century, with the increased market value of Priorat's wines (see chapter 4). Maurici and Josefina sent their son to study enology. Now in his early thirties, he works as an enologist for a local cooperative and has a small cellar producing his own high-end wine.

These processes of conversion were the rule of the land in Southern Catalonia. Yet the siphoning of nuclear money into farming did not always give rise to a virtuous cycle. In the first place, Maurici, the inheriting son of a middle *casa*, had a very different situation than that of poor *cases* and non-inheriting sons. The latter also used nuclear money to strengthen their resource bases, but becoming full-time farmers was largely out of their reach.

On the other hand, it is not coincidental that Maurici and Josefina are from Priorat. In Fatarella I could not find any case matching Maurici and Josefina's level of success. Two reasons explain this circumstance. First, Fatarella's nuclear workers did not experiment with new farming methods or sources of revenue; it seems reasonable to suggest that the villagers' high degree of involvement with the nuclear economy precluded it. Josep was the only person I spoke with in Fatarella who successfully engaged in new methods, and he never worked in the nuclear construction. The nuclear workers of Fatarella challenged the hegemony of the nuclear economy by being unruly workers and remaining antinuclear peasants. But despite their redirection of nuclear money into agrarian pursuits, at the level of practices, the nuclear economy was not so clearly defeated.

Second, the agrarian market played against the inhabitants of Fatarella. Most farmers from Fatarella reinvested their severance payments into hazelnuts, a crop that up to the mid-1980s was more profitable than olives and almonds and demanded less work than vineyards (which are also poorly adapted to the climate of Fatarella). Hazelnuts represented a risk-averse choice that allowed farmers to maintain one foot in agriculture and one outside of it. Although apparently careful enough, this choice proved to be full of risks.

From Village in Crisis to Ecological Village

The end of the nuclear construction coincided with Spain's entry into the European Union (EU) in 1986. It was a moment of profound redefinition in

European agrarian policies.[33] The protectionism that had characterized EU policy in the past was being replaced by a new liberalizing emphasis and a series of reforms—culminating with what is known as the McSharry reform of 1992—that favored the interests of large landowners over small and middle farmers. Mediterranean crops were most affected by the entry into the EU common market, with hazelnuts becoming the symbol of the deleterious effect on Catalan agriculture. Catalan hazelnuts were unable to compete with less costly Turkish hazelnuts.

No Southern Catalan village felt the "hazelnut crisis" as deeply as Fatarella, where plummeting agrarian prices coincided with the end of abundant jobs. The villagers' reaction was twofold. First, they became involved en masse in what the media called the "hazelnut wars," a series of mobilizations led by UP that demanded higher prices for hazelnuts and government help. The mobilizations became famous for their long duration, broad participation, and the direct action strategies employed. The hazelnut wars ended with a mediocre victory. In 1996, Southern Catalan hazelnut growers secured higher subsidies for the crop's production. To celebrate, UP organized a widely attended, daylong, national gathering in Fatarella. Nonetheless, the hazelnut wars were the last massive mobilization of UP, an organization that, by the turn of the century, had started to align with the interests of big farmers, thereby losing its hegemonic position and a part of its leadership (for instance, Josep) in the region.[34]

The municipality led the second response. After the local elections of 1991, won once again by the UP-backed independent list, the municipality officially declared Fatarella a "rural municipality in crisis." Led by Ferran, as mayor, and Fidel, as vice-mayor, the municipality created a socioeconomic commission to search for economic alternatives for the village: "We needed alternatives," says Ferran. "We had a terrible situation, with no nuclear jobs and the hazelnut crisis, which is where we had put the savings from *la nuclear*. There was a lot of pessimism, and we wanted to boost morale. So we tried to create new expectations around tourism, ecological alternatives, and renewable energy."

The project was part and parcel of a longer trajectory of antinuclear struggle, not only in that it continued the search for alternatives to the nuclear economy, but also because it came on the heels of the last outburst of antinuclear mobilization in Southern Catalonia. In 1989, Vandellòs I, the oldest nuclear plant in the region, suffered the worst nuclear accident in Spain's history. It was closed and dismantled. The accident triggered massive outcry in Southern Catalonia, with the antinuclear movement organizing a human chain of some forty kilometers linking Ascó and Vandellòs and demanding the closure of all nuclear plants. Largely in response to these mobilizations, the Spanish government passed a decree creating the so-called nuclear canon: the sixty-three Spanish municipalities situated within a ten-kilometer radius of a nuclear plant would receive an annual payment to compensate for the risks associated with nuclear energy.[35] In 1991, Fatarella received its first yearly payment of 20 million pesetas—one hundred and twenty thousand Euros—a sizable amount for a poor municipality. Fatarella aimed to *convert* that money, inject it into local reproductive circuits, and use it to generate economic alternatives to nuclear energy.

The municipality began a broad process of consultation with the villagers, hired a team of rural development consultants, and elaborated an economic development plan that was distributed as a book to every household in 1994.[36] The plan spelled a project of *autogestió*. Indeed, the development plan focused, in its own words, on promoting "a path of endogenous development" beyond agriculture. In fact, the plan clearly emerged from the realization that agriculture, on its own, would not sustain the livelihoods of the villagers. As economist Miren Etxezarreta, who was involved with the Southern Catalan antinuclear movement, said on entry to the EU, "agrarian development could not guarantee rural development."[37]

Despite the crisis of farming, the spirit of Fatarella's development plan had a familiar peasant flavor: "Our objective with the plan," says Fidel, "was to be autonomous." The plan proposed labeling the village an "ecological municipality" (*municipi ecològic*), a concept that was in line with the notion of endogenous development, as Ferran explains: "Ecological meant being respectful with nature, with our forests, our fields, and our stone wall terraces.... And quality of life, emphasizing that our [agricultural] products are tasty and healthy, that you can live here without much noise. Or building with stone, because it meant that you could take advantage of the resources that we have at hand. We were not ecological fanatics; we just wanted to find a balance with our environment. See, we have always had cisterns collecting rainwater, so this was the idea, not to be wasteful [*no malgastar*], to conserve."

The environmentalism Ferran describes fits with what Joan Martínez Alier calls *environmentalism of the poor*, "a convenient umbrella term for social concerns and for forms of social action based on a view of the environment as a source of livelihood."[38] For Ferran, "ecological" meant treating the environment with *cura* and trusting that the environment would then give back, providing resources for the economic development of the village.

In practice, the development plan focused on four kinds of actions that were developed in the following years. First came a series of measures oriented to boost Fatarella's image, attract tourism, and improve the quality of local products. The municipality created a small slaughterhouse to be used by the three butchers in the village, workshops promoting the virtues of organic agriculture, an annual "olive oil fair" in December, training sessions offering basic tourist skills to the population, and so forth. A second set of actions aimed to improve the quality of life in the village, primarily to prevent villagers working outside of the county from leaving it. To this end, the municipality created, among other things, a heavily subsidized daycare and a new youth social center with internet access.

A series of ecological initiatives formed the third leg of the development plan. These included the creation of a green water treatment system that expanded water availability to vegetable gardens and the construction of the "ecological house" (*casa ecològica*), a rural tourism hostel owned by the municipality. The hostel was built using a combination of green architecture—it runs entirely on the energy collected through a series of large rooftop solar panels—with traditional architectural techniques. Renewable energy was put at the service of endogenous development, showing that traditional rural mores were compatible with modern, green efficiency. Renewable energy was understood as ecological within the framework of the environmentalism of the poor: as a technology carefully integrated with the territory that could sustain local development.

Finally, a fourth set of measures oriented toward promoting vernacular architecture, at the private and public levels. The municipality engaged in successful efforts to revive drystone construction, almost lost among the younger generations, and in this way created an economic venue for the locals (who have a regional reputation as good builders). A series of public works—for instance, restoration of the town house and the extension of the football field—showcased drystone technique, with a few older local experts hired to perform the job and train young recruits. The municipality also passed new building regulations requiring the use of vernacular forms, techniques, and materials for private housing, guaranteeing a market for the skilled local labor.

The emphasis on vernacular architecture was a clear attempt to undo the effects that nuclear money had on the village's built environment. By establishing new building regulations and, more importantly, by engaging in a series of public works showcasing vernacular architecture, the municipality decoupled the association of stone with poverty and dependence. Vernacular architecture was reimagined as a source and demonstration of wealth and autonomy, strengthening the local resource base and forming a central piece in local development. This strategy generated a virtuous cycle. It gave local workers a de facto monopoly over construction activities in the village while creating a qualified and scarce labor force specialized in traditional techniques that were in high demand throughout the region. And, in beautifying the village, it spurred local émigrés to renovate their old houses now serving as summer residences. Vernacular architecture was the vehicle through which "urban money" was reinserted into the rural economy.

Taken together, the initiatives in Fatarella's development plan envisioned the generation of synergies between the creation of economic opportunities and the strengthening of the locals' belief in their own resources and possibilities.

Municipal leaders such as Fidel call it generating self-esteem: "You can only have an autonomous project if you believe in it, if you value what you have. We had to generate self-esteem, and I think we did it." In many ways this was the culmination of the peasant revitalization of the 1970s and the antinuclear countermovement of the 1980s—a project to bolster a peasant identity that was often expressed in the language of dignity. This connection was made clear in the inauguration of the ecological house, in 1998. In front of twenty-five hundred attendees, most from the counties of Priorat, Terra Alta, and Ribera, the municipal leaders presented the new facility as part of a project of resistance: "Initiatives like this one emerge from a firm and dignified will (*voluntat ferma i digna*) to live in our rural municipalities, to get the rest of the country to know about our products and life.... We are not an Indian reservation; we are normal people who struggle to live with dignity in our territory."[39] Living with dignity would involve living "ecologically," in the sense Ferran described: living in, from, and through the territory.

The various projects and plans yielded uneven results, combining some notable successes—such as the extension of drystone architecture, the reinforcement of the ties with the urban émigrés, and the olive oil fair—with equally notable fiascos, the most onerous probably being that the ecological house has failed to become the embryo of a new tourist sector. Even more troubling, Fatarella is still losing population. The younger generations, while voicing a widespread desire to stay in the area, often feel forced to leave in order to make a living. Nonetheless, the relevance and impact of the development plan of Fatarella must be assessed beyond its immediate failures and successes.

Most importantly, the development plan pioneered a new path in Southern Catalonia in a context where regional and national institutions had none. In 1995 the regional Catalan government approved a Territorial Plan whose aim was to assess Catalonia's geographical unevenness and equilibrate the territory. For this purpose, it defined a series of strategic development guidelines for all but three of the forty-one Catalan counties. Terra Alta and Priorat were simply identified as "stagnant agricultural areas with no prospect of improvement [*millora*]."[40] In other words, not only were they in dire straits, there was nothing that could be done to reverse their situation. Terra Alta and Priorat were redundant, fading into unprofitable obsolescence. Thus, while the government considered that these counties did not possess the resources to build their future—development could only come from outside and such a possibility was slim given their peripheral condition—Fatarella was elaborating a development plan that relied on endogenous resources, the ultimate outcome of the effort to build an antinuclear peasant moral economy.

The development path Fatarella pioneered would soon gain traction in Southern Catalonia. Several development initiatives, led by local governments and civil society, focused on the valorization of local resources through a combination of high-end agricultural products—fundamentally, but not solely, wine—and rural tourism. The local landscape, built over generations, was seen as the region's main economic resource and the basis of its future development.

As we will see in chapter 4, much of the debate around energy facilities in Southern Catalonia in the new century would come down to a confrontation between two development models. One model, expressed in Fatarella's development plan, would be based on the agricultural and landscape quality of the area, a continuation of the antinuclear and peasant impulse, stressing self-reliance and the autonomy of the region and the livelihoods of its people. The other, heralded by certain local elites and energy companies, was an extension of the morality of *la nuclear*: it presented the area as unable to self-develop, cursed by isolation, an inhospitable environment, and a peasant mentality attached to an unviable agriculture. In this view, local development was dependent on the arrival of new infrastructural projects, often energy facilities, generating labor demand and episodic boom-and-bust cycles.

Indeed, this tension between development models was seeded in the early years of Fatarella's development plan, as large-scale wind farm development began to scout the area. In the mid- to late 1990s, the municipal government was approached by several developers interested in building a series of wind farms in the village. At least initially, Ferran, Fidel, and the other members of the government received this interest with enthusiasm. Renewable energy would complete their development plan. Not only was it based on local resources—the abundant wind that local farmers had historically seen as a curse—but it was also in line with the ideas of a municipality that for two decades had complained about nuclear energy and extolled the virtues of renewable energy, putting forward modest but revealing initiatives such as the fully solar ecological house and renewable energy exhibits and workshops. However, as the mayor explains, the contacts with wind farm developers would soon trigger unease: "First came a person, in 1995 or 1996, saying that he wanted to put a wind farm, but he never came back. Later on another one came twice representing two different companies. And then a third one that made a clear offer, but two days later the second person came back and doubled that offer.... We soon realized that this was bigger and more complicated than we had imagined, and we needed more information, we needed to talk to other municipalities.... So we decided to postpone our decisions until the next legislature."

In the new century, the fields of Fatarella, in which the villagers had invested their nuclear money and dreams of autonomy, would become, as many other corners of Southern Catalonia, the object of intense wind energy development. The process would be pervaded with conflicts, as many saw only a new iteration of the continuous secular attacks against their autonomy and their dignity.

Notes

1. Ploeg (2013: 61).
2. Wolf (1969: 22).
3. For a discussion of this combination, see Gilabert (2017).
4. See also Rosen (2012).
5. Frigolé (1991).
6. On the youth association, see Associació de Joves de la Fatarella (1990).
7. Epstein (1958).
8. *La Cabana*, November 1, 1985, p. 12.
9. Sorribes and Grau Folch (1989: 182–184).
10. Carreras and Tafunell (2010: 231 and 381–389).
11. Logan (1985).
12. See, for instance, "Huelga indefinida en las obras de la central nuclear de Ascó," in *El País*, 16 May 1978; the newspaper claims that the strike affected four thousand workers.
13. Thompson (1971); Scott (1976).
14. Scott (1976: 157–192).
15. For recent reviews, see Edelman (2005), Browne (2009), Fassin (2009), and Palomera and Vetta (2016).
16. Palomera and Vetta (2016: 419).
17. Sider and Smith (1997).
18. See Mitchell (1990) for a critique of Scott's unilateral emphasis on the ideological aspects of domination and resistance.
19. Bourdieu (1977: 191). See also Scott (1985: 307–312).
20. *La Cabana*, February 19, 1987, p. 2.
21. These four covers correspond, respectively, to numbers 7 (February 1986), 13 (August 1986), 10 (May 1986), and 31 (March 1988) of *La Cabana*.
22. On the connection between procreation, reproduction and land barrenness, see Frigolé (1995: 58-83).
23. My take on extractive economies draws on: Bunker (1984), Coronil (1997), Ferry (2002), Ballard and Banks (2003), Wilk (2007), and Svampa (2013).
24. The synonymous *cuidado* is also often used. Ploeg (2013: 50–51) discusses the notion of *cura* in relation to the dairy farmers of Emilia in Italy. On farming methods in dryland Southern Catalonia, see Gros (2010).
25. Ploeg (2009: 274).
26. My use of the term "nuclear money" draws on Coronil (1997).
27. Ploeg (2009: 276).
28. Bloch and Parry (1989).
29. Ferry (2002: 343).
30. Ploeg (2009: 26). On these "operational logics," see also Shanin (1986).
31. Bohannan (1959).
32. Kopytoff (1986).
33. For an overview, see Etxezarreta et al. (1995); on the effects of subsequent reforms, see Majoral (2006) and Soler and Fernández (2015: 174–185).

34. Sociologist and agro-ecological activist Eduardo Sevilla Guzmán echoes this process: "UP was great, but now.... Yes, there are some dissident youngsters, but the rest ... they are businessmen!" (Sevilla Guzmán et al. 2008: 13).

35. Nowadays, the nuclear canon that Fatarella receives is on the order of three hundred thousand euros.

36. Ajuntament de la Fatarella (1994). The process initiated with this study continued in the following years, giving rise to the elaboration of an Agenda 21 for the village (Ajuntament de la Fatarella 2005).

37. Etxezarreta (1991).

38. Martínez Alier (2002: 40).

39. Quoted in Guiu (2008: 208).

40. Tarroja et al. (2003: 155). The third county was Garrigues, bordering Southern Catalonia to the north. The characterization of the Ribera was not too different: "stagnant industrial area" and "stagnant agricultural area evolving towards moderate growth."

Fig. 1. Village of Ascó with the chimney of the nuclear plants.

Fig. 2. "No more wind turbines," graffiti located on the road between Orta and Gandesa.

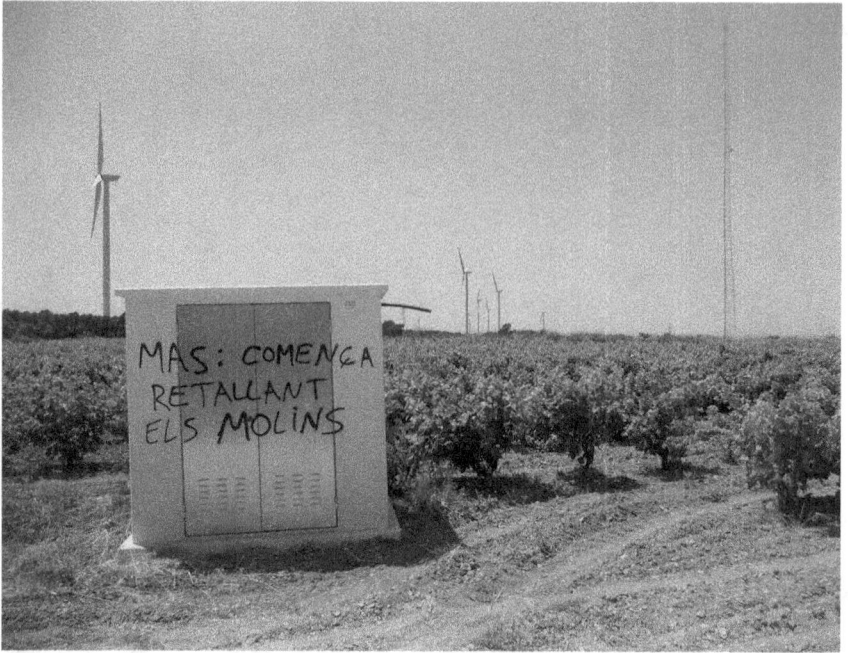

Fig. 3. "[Artur] Mas: start your cuts with the turbines." The graffiti, located in Vilalba, makes reference to the austerity cuts introduced and defended by the Catalan government during the time that Artur Mas was its president (2010–2016).

Fig. 4. Demonstration against the nuclear cemetery in Ascó (2010).

Fig. 5. Dry-stone *cabana* with a wind turbine on the background.

Fig. 6. Substation on the path to "the end of the world."

Fig. 7. Image of Fatarella's agrarian hinterland, with Corbera's wind farm on the background.

Fig. 8. Transport of wind turbine blade, with farmers from Vilalba returning from harvesting grapes waiting by the side of the road.

4   Southern Revolt

Capitalism does not have an ecological regime; it is an ecological regime.
—Jason Moore, *Capitalism in the Web of Life*

We are sister struggles, we all fight against a development model that makes us peripheral: a pantry for water and energy, and a waste dump for what the country does not want.
—Susanna, activist of the PDE

On FEBRUARY 4, 2001, the streets of Móra, the small commercial capital of the three northern counties of Southern Catalonia—Priorat, Ribera, and Terra Alta—hosted the largest demonstration that has ever taken place in the area. In front of twenty-five thousand peaceful demonstrators, a simple banner read: "Stop aggressions to the territory." It was signed "The Platforms," a term that in Spain identifies local civil organizations convened to oppose a specific localized development and operating through an assembly-based, nonhierarchical structure. The four self-identified "Southern Platforms" that organized the Móra demonstration formed in the previous two years in response to three kinds of infrastructure projects. The Platforms of Terra Alta and Priorat opposed the proposal of the Catalan government that positioned the two counties at the center of wind farm development. The Platform of Ribera opposed several projected waste and energy facilities, most notably, Enron's plan to build a natural gas combined-cycle power plant in Móra. And the Southern Catalan section of the nationwide Platform in the Defense of the Ebro (PDE) opposed the National Hydrological Plan (PHN, in its Spanish acronym), a public plan of hydraulic infrastructures that hinged on the transfer of water from the Ebro River to Barcelona and the peninsular southeast.

The Móra demonstration is widely seen as the inaugural for an unprecedented cycle of mobilization, popularly known as the Revolt of the Ebro or the Southern Revolt. This chapter examines this long cycle of mobilization from a specific angle: opposition to wind energy. Wind was no minor issue in the Southern Revolt; two of the four platforms organizing the Móra demonstration were fundamentally devoted to opposing what they called "wind farm overcrowding" (*massificació eólica*). The expression is important, because it makes explicit that the platforms were not against wind energy, but rather its concentration in an area that already produces 70 percent of the electricity generated in Catalonia though it represents only 3 percent of its population.[1] Indeed, the conflicts that composed the Southern Revolt should be read as what political ecologists and ecological economists call "ecological distribution conflicts," struggles to redress the uneven distribution of environmental burdens and benefits.[2] The location of wind farms in Southern Catalonia was perceived as unjust within a broader historical legacy of peripheralization and unequal exchange at both the economic and ecological levels. This peripheral condition united the different struggles that coalesced in the Móra demonstration, as the second epigraph of the chapter shows.

The revolt was experienced as a moment when a long history of peripheralization crystallized into historical consciousness. Finally, the South had a voice. And this voice, by and large, spoke in the language of dignity—the cultural code and mobilization instrument through which Southern Catalans attempted to wrest control over the decisions that affected their territory and livelihoods. Manolo Tomàs, the charismatic leader of the PDE, expressed it thus: "The will that we Southern Catalan people are expressing is the will to be and to exist. We call it Southern Catalan dignity.... What we are questioning is, in the end, the current territorial model. The concentration of resources and infrastructures in big cities condemns us to poverty and the depopulation of peripheral areas, and this must stop."[3] The appeal to dignity thus points, in Narotzky's words, "to the feeling that the grounds on which certain principles of inequality were tolerated are no longer there."[4]

As a whole, the Southern Revolt emerged as a radical systemic critique denouncing patterns of accumulation and undemocratic political practices that had reproduced the region's peripheral condition. Through this critique, the Southern Revolt uncovered Spanish capitalism as an *ecological regime* built on an intense mobilization of space and nature and the systemic reproduction of unevenness.

The Beginnings: Disorientation and Disorder

In the mid-1990s, coinciding with the beginning of the expansion of wind energy in Spain, a host of small- and medium-sized wind developers started to visit and offer deals to the mayors and landowners of Southern Catalonia, especially in the counties of Terra Alta and Priorat. This gave rise to a period, roughly between 1996 and 1998, marked by feelings of disorientation and a generalized perception of disorder. Among ordinary citizens, disorientation was fueled by their lack of information about and experience with the sector, as well as by the contradictory responses of local mayors, who faced a cascade of wind farm projects (by 1997, as many as thirteen had been announced in Terra Alta and nine more in Priorat).

As a general rule, conservative mayors were unabashedly enthusiastic about the prospects of wind farm development, celebrating it as a new Eldorado. In successive years, as popular opposition and bureaucratic burdens forced the delay or withdrawal of several wind farm projects, these mayors would express their frustration in a new mantra: "We want to live off wind."[5]

In contrast, left-wing municipal governments reacted cautiously, considering the process complex and confusing. They lamented that the Catalan government was not providing clearer guidelines and advice, and they called for a county-level agreement to allow the municipalities to pool resources and design a comprehensive plan for wind development in their region. Bernat, councilman of Fatarella between 1999 and 2003, recalls one of these attempts: "We had a meeting with all the other mayors, and from our municipality we proposed that we make a decision as a whole: How many turbines [should there be] in the county? Where? What do we want in exchange?"

Several alternatives were formulated, but this approach never gathered a broad consensus: "No municipality wanted to renounce its own project," says Bernat, "No one set guidelines. It was a total disaster, and it was the Catalan government's fault. They sent the developers here, and this was a rush."

While in Spain electric regulation—including, for instance, subsidies to renewable energy—is in the hands of the central government, wind energy development is largely regulated by each regional administration (or *Comunidad Autónoma*), which enforces planning guidelines and ultimately has to authorize installation. Yet by the time that wind energy started to arrive in Southern Catalonia, the only legal document that the Catalan administration had ever crafted was a Wind Atlas. Issued in the late 1980s, the document was a mere cartographic description of wind speeds in the Catalan territory. As such, the Wind Atlas did nothing to organize wind energy development in the region, but it was a crucial tool informing the developers' decisions on where to site their projects.

The Atlas turned what had been a free-access good with no value into *abstract social nature* with potential market value.[6] We may observe here a clear instance of the *alienation* of nature that stands at the basis of capitalist modernity: posited as abstract and external to the world, the environment is fragmented into stand-alone resources that can be appropriated and subjected to price calculations. By mapping and quantifying wind, the Wind Atlas was a knowledge-producing device that constituted a necessary first step for the construction of a new resource and a new commodity frontier. The wind rush could proceed.

The sense of disorientation among Southern Catalans was crucially reinforced by the opacity of the energy sector and the deals that it was offering during this boom. A good example of this opacity occurred between 1994 and 1996 in Pradell, a municipality of Priorat. Mercè, an émigré from the village who returned in the 1990s, explains: "Everyone was saying 'there's a crazy man buying the mountains. What a stupid thing to do, these lands are all stone, they have no water.' Of course, this was just a middleman, and no one knew that it was for a wind farm—well, two in fact. And then, in 1996, they registered the wind farm, and it was chaos when people found out. But the companies already had the land, and the municipality had to cut a terrible deal with them."

Pradell's wind farms went into operation at the turn of the century. Although Pradell's case is in some respects quite exceptional—wind developers rarely buy the terrains or are so secretive—confusing information was a recurrent characteristic of the deals being offered. Consider, for instance, the experience of Emili, a farmer from Fatarella, who was approached by a wind developer in 1998:

> So one day, I was with my father in that field that we have in Vilalba, and then some people from Barcelona showed up. They told us that they were part of a company called Zart. They were trying to look friendly: "Is it very windy around here?" You see, what a question! Anyway, they asked me if they could install a wind meter. At that time I had never seen one of these meters yet. We listened to them. They gave us a letter explaining the details of what they wanted to do in our field, and said that we would later on receive a contract to sign. I mean, we weren't very interested, mind you, but once we saw the contract! Look, it was full of technicalities and small print. But as if that was not enough the contract had the stamp of Enron. Can you believe it? Of course, we didn't sign it, but they ended up making a deal with the neighbor.

Emili's confusion had its origin in the complicated deals going on in the wind sector during this early stage, a point to which I will return in chapter 5. A related source of confusion was the proliferation of opportunistic brokers trying to secure favorable locations. In the jargon of the sector, this was known as "getting a REPE," an acronym that stands for Special Regime for Electricity Producers, the legal category that allows renewable energy producers to obtain subsidies. Typically, after brokering a deal with a mayor, these opportunists, who often possessed ties in the regional administration, would register their project, and, once they obtained the REPE status, they would sell their rights to a legitimate wind company, thus making remarkable profits while committing almost no capital.

The proliferation of wind farm projects being announced by the local press and the pervasiveness of opportunistic practices, together with the almost complete absence at that time of regulations around wind farm siting, led many citizens to view wind energy with fear and suspicion. *La Serena*, a recently created monthly magazine from Gandesa, the capital of Terra Alta, became the forum in which to voice these fears. The editorial in its first issue, from 1997, summarizes:

> [We in Southern Catalonia] have always helped the rest of Catalonia and Europe when it comes to producing dirty energy. Now we see the construction of a wind farm in Baix Ebre. This would be fine were it not that the European

Union guidelines recommend that new energy-producing units be located near energy-consuming centers. And here in Southern Catalonia or in Terra Alta, there isn't much consumption! ... We are in favor of wind energy, but only if it means that, more or less quickly or slowly, it can replace nuclear and thermal power plants.... As for the wind farms, their development should be rational and balanced. It should be said very clearly that some places are fine, but others are not! In other words, there should be a consensus within civil society.[7]

In its attention to the imbalance between electricity production and consumption, lack of planning, the silence around nuclear power, and the fear that wind farm development may disrupt traditional and emerging economic activities, the editorial flagged the main issues that would remain at the heart of the wind energy debate. The group of middle-class, liberal professionals gathered around *La Serena* responded to these fears by spearheading, in 1999, the creation of the Platform for Terra Alta—initially named Platform for the Defense of the Natural, Historical and Cultural Patrimony of Terra Alta—following the example of the Platform for Priorat, created a few months earlier with similar objectives. The founding manifesto demanded "an ordered development of wind energy, respectful of the landscape and of natural and historical resources." The emergence of the platforms signaled the formal beginning of the conflict around wind energy in Southern Catalonia.

Territorial Value Conflicts

In 1998, the Catalan government approved the Plan for the Development of Wind Energy, popularly known as the Wind Map. The Wind Map was supposed to put an end to the wind rush of the preceding years, providing a regulatory framework for the development of wind energy in Catalan territory. Nonetheless, the Wind Map was largely devoid of regulatory content; indeed, it was little more than a map specifying seventy-five locations deemed suitable for siting wind farms. As geographer Sergi Saladié argued, the Wind Map did not really regulate wind farm development, but gave guidelines to wind developers while providing a legal foundation for their existing projects.[8] The Wind Map projected the installation of 1,329 MWh of electrical capacity from wind by 2010 (12% of the electricity consumed in Catalonia), and the overwhelming majority of the locations it marked were in the mountain chains of Southern Catalonia, where wind developers had already begun focusing their efforts.

If the Atlas made wind into *abstract social nature*, the Wind Map tied access to the wind resource to the control of Southern Catalan land, transforming the latter into what Henri Lefebvre would call a dominated, homogenized *abstract space*, adjusted to the extraction of energy and profit.[9] The Wind Map established a chain of value relations, weaving together abstract nature, abstract space, and abstract energy and subjecting all these elements to capital's law of value. Along with the production of a valuable abstract nature, abstract space had to be produced as valueless, in a process whose logic is described by Elmar Altvater: "Valorization—or, in other words, extension of the market's formal rationality and scarcity principle to previously free resources—always entails a largely hidden definition of the 'nonvalorizable' or 'valueless' object whose destruction is permitted. The price of 'valorization' is thus the ruin of that which has no economic value."[10]

In effect, the Wind Map offered a picture of the country that corresponded with the one depicted by the Catalan Territorial Plan in 1995. Wind energy development was to be concentrated in those areas that the Territorial Plan had described as having "no prospect of improvement," that is, unable to develop by themselves. From the perspective of the Catalan government, wind energy production represented the "best use" for the lands of Southern Catalonia, the best way to make a profit and bring development to the region. This also meant that the Wind Map implicitly associated wind development areas with poverty, a lack of entrepreneurship, and, crucially, low territorial value. Wind developers used environmentally deterministic logic to support that logic, thus obscuring the political economic production of unevenness that sustained it. For instance, Ernest, a wind developer with interests in Terra Alta and other Catalan counties, told me, "Wherever I go [to develop] I find backwardness, because wherever there is wind there is poverty, because wind dries up the land, so as wind developers we always end up going to poor regions lacking opportunities." The Southern Catalan wind conflict would emerge as a reaction to the vision of Southern Catalonia as a valueless territory and a resource frontier, that is to say, as an effort to dignify that territory.

Development Models: Bowls of Lentils and Wineries

The conflict around wind farm development emerged as a political clash between supporters of two different economic models. The groups attached very different value to the local resources and territory. The first, endorsed by the platforms, emphasized the value of local resources and imagined a future in which traditional agriculture would coexist with rural tourism and high-quality wine (and olive oil) production. The second model relied on the arrival of infrastructural projects to bring jobs, leases, and taxes and was vehemently defended by those mayors who "wanted to live off wind." Almost all in this latter group belonged to the Catalan nationalist conservative party CiU (Convergència i Unió), which had controlled the Catalan government without interruption since the early 1980s.

The support for wind energy, at least in political circles, was based on reading the region's environment as valueless and unproductive, the ultimate explanation for its underdevelopment. No one ever gave me so stark an expression of the pro-wind energy vision of the land as miserable as did Gustau, former long-time mayor of Móra. An engineer professionally linked to the electric sector and a veteran power broker in the region, Gustau, like the vast majority of conservative mayors, had historically supported nuclear energy. Even if he welcomed the arrival of wind farm projects to the region, he was skeptical: "I don't know if you can have a whole electricity system based on wind, I doubt it. I mean, I am not the biggest fan, but I am happy that they build them here. In fact, it's all energy, right?" As we continued our conversation, it became clear that his enthusiasm about any kind of "energy" was built on a profound dislike of the region's environment: "What do we have here? Do we have an agrarian land with a series of orographic characteristics allowing the development of agro-food industries? No. Do we have beaches or mountains for skiing? No. Do we have oil? No. What do we have? We have wind, in certain places. We have gas pipelines. And we have a river, that if there is something that we use it for it is to refrigerate the nuclear plants. We haven't been able to find another use for it. If we didn't have the nuclear plants this would be a desert."

Behind Gustau's words emerges a vision of nature as valuable only insofar as it can be dominated and converted into a bulk resource for a growth-oriented economic metabolism, whether in the form of mass tourism, electricity, water, or agro-business. Nonetheless, during the 1990s, a different approach to local economic development was starting to gain traction. This approach conceived the local land and landscape as a central resource of a different kind, pointing toward the idea of endogenous development that the municipality of Fatarella and Unió de Pagesos had pioneered in the early 1990s. Mireia, who directed the office of economic development of Terra Alta, explains:

> Since the mid-1990s our idea has always been the same: fomenting tourism, irrigation, quality wine. We were lucky because we had funds from the LEADER [a EU program for the development of rural areas] and that helped to push some initiatives. We will not live just from tourism, or just from agriculture, or just from industry: it has to be a bit of everything, a diversified, sustainable economy, living a little bit from everything. As peasant households have always done, haven't they? This is the idea. We don't want to make a county just waiting for a large corporation from elsewhere to come and save us. This is not the way. The value of this county is its own value [*valor propi*], with local companies, because the companies that come from elsewhere end up leaving and this creates a problem.

The founders of the platforms, mostly middle-class professionals and small business owners linked to the emerging tourism and wine sectors, feared that developers' promise of easy yet uncertain revenues would hamper their new economic orientation. In biblical language, they would say that local mayors were "selling the region for a bowl of lentils." The founding manifesto of the Platform of Terra Alta put it thus: "The indiscriminate construction of wind power plants [*centrals eòliques*] could negatively affect traditional agrarian activities, the promotion of local wines, the development of tourism and leisure and our capacity and strength to move forward our own initiatives and projects [*iniciatives i projectes propis*]."

The mistrust of wind energy was not based on its incompatibility with local development at a functional level, since in principle wind farms are technically and legally compatible with rural tourism or wine production. Rather, the mistrust stemmed from the fear that wind development might discourage the project of endogenous development by eroding the local population's belief in the region and the project; if wind development reinforced Gustau's idea that the land was good for little else, what hope would there be for projects—such as tourism and wine production—that relied on positing the region as valuable? As we saw in the last chapter, the wager for a local development path was a wager for strengthening the dignity and self-esteem of the population, understood as the population's belief in its own capacities and resources. In this light, wind energy was seen as embodying a return to a position of passivity and dependence.

Instead, endogenous local development was seen as an opportunity to reverse the legacy of rural exodus and the tendency of younger generations to emigrate in search of life opportunities. Even more important, since the opposition against "wind farm overcrowding" was understood as a defense of that path of endogenous development, the mobilization against it worked in itself as a catalyst for the return of emigrated professionals. Núria, the leader of the Platform of Priorat, exemplifies this process. She was born and raised in Molar, a small village in Priorat, where she has deep kinship networks. She is well known for her activist past, especially for her involvement in the last major episode of the antinuclear struggle, around the 1989 accident in Vandellòs. Initially, she watched the arrival of wind energy from her residence in Barcelona, where she worked as a philologist: "At first, coming from the antinuclear movement, I thought that wind energy was a good thing, but then you start seeing the way they go about, the places where they want to put wind farms, and you say oops!" Núria had also been thinking about the possibility of returning to Priorat for a long time, but never found the right moment. The mobilization against wind energy generated the circumstances: "With the platform I got more involved, spending more and more time here, and I ended up coming to live here. For a year or so I was commuting to Barcelona every day but I already had the idea of doing something [different]." In 2000, she opened a rural tourism establishment. Núria was not alone; rather, she formed part of a small but budding rural cultural and ecological tourist sector.

Yet, the critical impulse behind the platforms did not come from the emerging tourist sector, but from the wine sector. Wine had long formed the backbone of the local economy of Terra Alta and Priorat, but in the 1990s it was experiencing crucial transformations. Antònia, a leading voice behind *La Serena* and the Platform for Terra Alta, is heir to extensive agricultural property. Her family had historically processed wine and sold it in bulk, but in 1994 she decided to reorient the family business and created a winery. She believed "this was the path to follow; we had to create more value if we wanted new generations to continue to be invested in the area." In Priorat, similarly, local agrarian schools and some key actors within the wine sector played a crucial role in the mobilization against wind development.

Yet, there was a key difference between the two counties that would have dramatic consequences. While in Terra Alta the wine sector was completely dominated by local cooperatives and wineries, in Priorat these local actors coexisted with a handful—about half a dozen by the late 1990s—of wine entrepreneurs from other Spanish and Catalan regions. They were known as the "young sons" in reference to the fact that they were the non-inheriting sons of well-known high-end winemakers who saw the wines of Priorat as an opportunity to build their own names in the world of wine. These "young sons" were among the earliest supporters of the Platform of Priorat, bringing with them the powerful DOs (Denominacions d'Origen, local wine regulatory councils) and ultimately the rest of the wineries. By contrast, Antònia was unable to bring the less dynamic wine sector of Terra Alta to join the positions of the platform: "At some point most people thought that the platform was just me!"

Popular knowledge in Priorat has it that Pau was the key agent in bringing the county's wine sector to the "antiwind" side. The son of a middle *casa* of Priorat, he studied enology in Barcelona, planning his return to his county. In the mid-1990s he became an enologist for one of the "young sons," a job that he combined with a teaching position in the newly created School of Enology of Priorat. Together, his positions gave Pau a great deal of influence. He spread his views among younger generations and encouraged his employer to take a public stance against wind energy development in the region:

> We [he and his employer] saw it very clearly from the beginning. Look, when you do a high quality wine you are also selling an environment, a landscape, a certain way of doing things. And what we had on the table was a totally industrial form of wind development, incommensurate with our environment [*no raonada amb l'entorn*].... This is why we opposed it and we tried to defend our standard, which is a standard of progress, of wanting to be who we are. We have to demonstrate that we can be one of the best wines of the world. This is good image and prestige for Catalonia, and it is an industry that cannot be delocalized.

Ultimately, Priorat would be far more successful in defeating plans to construct wind farms in the county. This success would prove to be a burden on neighboring counties like Terra Alta that lacked a strong coalition to deter wind development. Indeed, the pressure on the agricultural lands of these "low value" counties would only increase as activists also won the battle to restrict wind farms in the region's mountain chains.

Sacred Places: Relatedness, Value, and the Environment

By linking the expansion of wind energy to specific locations, the Wind Map did not only provide guidelines and support to developers, it also inadvertently helped establish the activists' strategy: the most effective way to paralyze wind development was to prevent access to these locations. The production of wind as an economically valuable resource was premised on a process of objectification that disembedded the environment from social relations. Southern Catalan activists responded by engaging in what Hornborg terms a "struggle for relatedness."[11] Mobilizing a plethora of *languages of valuation*, the activists claimed the nonequivalence of Southern Catalan lands, asserting their *intrinsic value*— which interestingly enough is Kant's definition of dignity.[12] This process became manifest in the platforms' two main struggles during this period.

Terra Alta devoted the lion's share of its energies to preventing the installation of wind farms in the mountains of Pàndols-Cavalls. The value of the mountains stemmed from historical memory: Cavalls and Pàndols, the central stage of the Battle of the Ebro, evoke mythic echoes of the Civil War among veterans, academics, and the local population. In the decades after the war, Cavalls-Pàndols were filled with humble commemoration sites, as well as trenches, war rubble, and human bones. Veterans, particularly those on the losing side, including International Brigades, returned to pay respects. Thus, in its founding manifesto of 1999, the Platform of Terra Alta called the area "a sacred land for people from all over the world." In April 2000, the two platforms, with the support of a coalition of historians and academics and a large number of civil associations from the region, organized a gathering in Cavalls, attended by two hundred people, with the motto: "Yes to wind energy. No to the destruction of spaces of natural and cultural interest."[13]

The mountain of Montsant, for its part, became the battleground of the wind energy conflict in Priorat. This central orographic landmark hosts the ruins of an old Carthusian monastery that gives its name to the county. Locals saw the several wind farm projects in Montsant as a humiliation, further evidence of their degradation and peripheral condition. Using a formulation that I heard on multiple occasions, Núria, the leader of the Platform of Priorat, explains her initial reaction: "A wind farm in Montsant? In Montsant? Are you nuts? Can you imagine that someone said that she wants to put a wind farm in Montserrat? You can't, right? They wouldn't let it happen, the powers that be or the people or whoever would prevent it, right? So this is the same: the Montsant is our Montserrat."

The comparison with Montserrat, a mountain and an abbey situated in central Catalonia, not far from Barcelona, widely considered the sacred mountain of the Catalan nation, is highly revealing.

The Wind Map's categorization of Montsant and Pàndols-Cavalls as suitable locations for wind energy development was based on their wind regime, setting aside any other characteristic of these mountain ranges. In claiming their sacred character, the platforms attempted to disrupt the value relations implicit in the Wind Map by establishing an incommensurable *value regime*.[14] The platforms insisted that these mountains were inalienable, places whose "aura"—to use Benjamin's expression—was not and could not be for sale, for it could not be quantified using the metrics of economic valuation.[15] Yet the platforms' action was not devoid of economic purpose.[16] Both Montsant and Pàndols-Cavalls were central pieces of the model of self-centered development that the platforms defended. Pàndols-Cavalls was to be the distinctive element allowing Terra Alta to develop its own brand of cultural-cum-rural tourism, and Montsant, the iconic visual reference of Priorat, would be evoked by winemakers who claimed proudly that "inside every bottle of wine there is a piece of the landscape of Priorat."[17]

The platforms' success in impeding the installation of wind farms in Montsant and Pàndols-Cavalls was crucially linked to their collaboration with Gepec, a naturalist organization headquartered in Reus with deep ties in the region. Already by 1998, Gepec had started filing appeals against nearly all wind farm projects in the region. The organizers' main argument was that wind farms threatened the fragile ecosystems of the Southern Catalan mountains—which had historically been largely uninhabited and undeveloped. Specifically, the installations would threaten certain species of endangered birds. This argument was well supported by environmental legislation, for the majority of Southern Catalan mountainous spaces (covering around a third of the region's area) are included in Natura 2000, EU's network of nature protection areas. Well furnished with environmental assessors and legal counselors and possessing long experience in juridical battles, Gepec's strategy pitted the Catalan Ministry of Industry—in charge of wind farm development—and the Ministry of Environment—in charge of environmental protection—against each other in a battle that, as government officials admit privately, is still ongoing.

In practice, Gepec's appeals paralyzed a large proportion of wind farm projects. Developers were forced to undertake costly, time-consuming assessments of environmental impact that added a new burden to the already complicated bureaucratic process of obtaining authorization for the operation of a wind farm. For almost a decade, there were only three wind farms, including Pradell's, standing in Southern Catalonia, all finished at the turn of the century.[18] This paralysis triggered an angry and well-publicized reaction from the mainstream Spanish environmentalist organizations, fully supportive of renewable energy development.[19] As a very well known environmental activist with deep ties in the wind industry once told me, "These groups like the Gepec are not environmentalists, they aren't. They are preservationists, and this is not environmentalism."

The conflict between these two kinds of organizations—naturalists and mainstream environmentalists—can be expressed as a conflict between two opposing understandings of the "environment." Yet this stark opposition rests on a shared epistemological position that posits the environment as an external entity, therefore reproducing the division between nature and society that is central to Western modernity's ontological foundation.[20] The expertise of these two groups rests on that externality. Tim Ingold calls it *global perspective*.[21] Taking the environment as an object to be managed, the global perspective of the environmentalists conceives of wind energy as a technological mechanism to fix "nature." The law and the state support this argument, which is ultimately behind renewable energy subsidies. Naturalist groups like Gepec mobilize a variant of the global perspective that Ingold calls *ecocentrism*: "an attitude which credits the world of nature … with an intrinsic value quite independently of the purposes and activities, and even the presence, of human beings."[22] Indeed, for Gepec the value of the mountains of Southern Catalonia stemmed from the absence of human activity, and opposition to wind farms was oriented toward preserving that supposedly "pristine nature." The success of Gepec's strategy was largely based on the fact that the state also supported this ecocentric vision, which is easily detectable behind legal frameworks such as Natura 2000.

In one respect, the alliance that the platforms established with Gepec made a lot of sense. Both kinds of organizations opposed the proliferation of wind farms in the region, and they both based their strategy on claiming the inalienable character of the sites where wind farms were to be installed. Yet, in another respect this was an uneasy alliance between two worlds that, to use Marilyn Strathern's expression, were only "partially connected."[23] Indeed, the defense of the "local environment" constituted an instance of what Marisol de la Cadena, drawing on Viveiros de Castro, calls an *equivocation*: homonymous terms that refer to things that are not the same.[24] Gepec's ecocentric vision of Southern Catalan mountainous ecosystems as possessing an intrinsic natural value was quite alien to ordinary Southern Catalans. Particularly among those closest to farming activities, valuable nature is not "pristine nature," but nature that is humanized, epitomized by household agriculture. Southern Catalans do not value nature as an object to be managed and protected from outside, but as an environment to be taken care of and co-constructed from within. Pau explains this notion of nature:

> It is impossible to make a profit from almonds. We keep the almond trees to have [the field] clean, because if you don't have almond [trees] it would be dirty, and the same with the olive tree. Without them [the field] would get lost [*perdut*,

i.e., without direction]: we would have all kinds of weeds and this would be the jungle. Which is what happens to the forest when you don't manage it: the fire comes and everything starts burning and everything to hell. Here we are managing the biomass: an almond tree fixes CO_2, from time to time I plow it, the same with grass. There is care, an ordered management, and this retains the balance, fixes the value. But a forest doesn't, it doesn't retain anything, it is a disorder. What we demand is ultimately order, like everyone else, a certain order, nothing else. And nature also wants that order, an order that allows animals to live. But a forest is total disorder; the winner is not the strongest or the smartest, but the lucky one. The forest is the domain of luck [*la garriga és la sort*]. And what we wish for ourselves is not luck; what we want is order.

Gepec's ecocentric perspective was only partially connected to the peasant notion of nature described by Pau, which was closest to the *environmentalism of the poor* that, as we saw in chapter 3, is primarily concerned with the environment as a source of livelihood. The model of endogenous development defended by the platforms was linked to this form of environmentalism, which rarely expresses itself in environmental terms. The platforms' struggle for relatedness was a struggle to control the uses to which the land was subjected, to have a voice and decide about their own future, to channel their land toward the reproduction of local livelihoods.

Yet, it was not the platforms' brand of environmentalism but rather Gepec's ecocentrism that was more successful in paralyzing wind farm development and, ultimately, affecting public policy. This had two undesired consequences. On the one hand, the government began to force developers to commission "environmental impact assessments," carried out by environmental consulting firms. The decision about the environmental impact of a wind farm was to be determined by experts possessing a global perspective, thus rendering local environmental knowledge largely irrelevant. On the other hand, both wind developers and the administration grew progressively aware of the risks of siting wind farms in forested, lowly anthropized, mountainous areas that fit the ecocentric definition of valuable nature. As a result, wind developers directed their attention to the Southern Catalan agricultural landscape, thus displacing wind farm development toward the areas that Southern Catalans saw as their main collective and personal patrimony. As we will later see, as the paths of Priorat and Terra Alta began to diverge, the agricultural lands of the latter would become key sites for wind energy development.

The Southern Revolt

In the fall of 2000, the provisional approval of the National Hydrological Plan (PHN) and the parallel announcement of Enron's project to build a power plant in Móra added to the wind conflict and unleashed an unprecedented cycle of social mobilization that came to be known as the Southern Revolt. While in many ways it is ongoing, the revolt peaked between 2000 and 2004, coinciding with a broader cycle of political unrest throughout the country (including anti-war demonstrations, the antiglobalization movement, and protests around the *Prestige* oil spill in Galicia), as well as the mushrooming of local environmental and territorial conflicts all over Catalonia.

During those years, the "Southern Platforms" sustained an intense campaign of assemblies, marches, legal actions, town-hall meetings, and contacts with political leaders. The platforms discussed and elaborated alternatives to the projected infrastructures. Despite their small number of active members—in some cases no more than a dozen individuals—the platforms enjoyed significant support in the region, where they achieved a hegemonic position and connected with a diverse social base, as well as in the rest of Catalonia. In March 2001, three hundred thousand people—far more than the total population of Southern Catalonia—attended a demonstration organized by the platforms in Barcelona. This is especially remarkable considering that the main trigger of the revolt, the PHN, included the transfer of water to Barcelona's metropolitan region.

The impact of the Southern Revolt was deep and durable. Among its achievements and successes, four stand out. First, it signaled the emergence of a new political actor that shifted the balance of political forces within Southern Catalonia and, by extension, the rest of Catalonia. Second, it played a crucial role in the withdrawal of some key infrastructure projects. Third, the southern platforms generated credible, radical alternatives to the projected infrastructures, which they named the New Culture of Water and the New Culture of Energy. Finally, through its radical practice, the Southern Revolt emerged as the first broad popular critique of the patterns of accumulation dominating Spain's political economy at the turn of the century, visualizing an ecological regime that sustained accumulation. Let me begin with the first two points.

Right after the provisional approval of the PHN, the car bumpers of Southern Catalonia were filled with stickers declaring "The river is life," "No to the transfers," "The South says enough," or "We are also Catalans."[25] Proudly using local dialect, all these messages conveyed the mix of outrage and pride—indignation and dignity—that was the hallmark of the movement. Perhaps the bumper sticker that best captured the feeling of the Southern Revolt was "We are not a reservation." The complaint was not new. In the winter of 2000, the platforms of Terra Alta and Priorat had organized an act of protest against the Catalan president, who was on an institutional visit to the counties; on his arrival in the village of Poboleda, two hundred protesters dressed as Native Americans booed him. Southern Catalans felt that they were being treated as second-class citizens, dominated subjects in their own land. From their perspective, the region was an "internal colony" and thus the target for a series of projects that, in the view of Southern Catalans, undermined their livelihood projects and development potential. Valerià, a long-time, highly politicized antinuclear activist and the grandson of anarchist collectivizers from Terra Alta, puts it boldly: "It's been long decided that we are surplus, they're pushing us out. A handful of alien companies have decided that this country has to be destroyed. And this is the result of a historical process: eliminating the people from here while making profits. I am not a Christian, but there is a sentence that says 'Thou shall defend the land of your parents,' and this is our house. That's all: this is where our ancestors were."

The Revolt of the South was, however, not only a struggle against "outside aggressions," it was also an internal battle along clear-cut political lines: the revolt emerged as a force of change on the left that opposed the status quo, personified in those holding positions of privilege and authority, who had historically been supportive of the arrival of infrastructure projects. While the platforms have been able to stop many of these projects, their main success was to transform the political landscape of the region, where the right is no longer hegemonic.

Combined pressure from the southern platforms yielded early, tangible results in the Catalan parliament. On March 8, 2001, just a month after the Móra demonstration that we saw in the chapter's opening, the Catalan parliament discussed a joint motion against the infrastructural projects in Southern Catalonia. In one single session, Parliament withdrew the Wind Map (committing the government to present a modified version later that year), and dropped its support for the Enron power plant and for the PHN. This last decision was largely symbolic, for the approval of the PHN depended on the Spanish parliament, where the ruling Partido Popular (PP), amid widespread partisan and public opposition, used a steamrolling majority to pass the project in 2002.

The surprising chain of events that took place in the Catalan parliament in 2001 was made possible by the growing tensions and disagreements between the two conservative parties—CiU and PP—that dominated the complex and fragmented body.[26] These tensions were a direct consequence of the Southern Revolt, which forced CiU, initially supportive of the PHN, to shift its position once it became obvious that this support was dramatically eroding its popularity. Yet CiU's shift did little to reverse its accelerated weakening position in Southern Catalonia, and in 2003, it experienced its first-ever electoral defeat. The new Catalan government was formed by a coalition of the three left-wing parties that had presented the joint motion in 2001.

Núria admits that the concatenation of parliamentary events in March 2001 was "a fluke [*carambola*], a lucky fluke, yes." Yet she adds, "But we [the southern platforms] wrote the script." Fidel, the antinuclear activist and former vicemayor of Fatarella, agrees, quoting passages from the motion by heart. The

Southern Catalan social-democratic parliamentarian Xavier Sabaté presented the motion in an intervention that clearly communicated the concerns of the southern platforms:

> The government does nothing about the problems of Southern Catalonia; it has no model for that territory. The neglect of the South is real.... People say [it] in plain language: "Our water allows some to grow more and us less." Energy production facilities generate few jobs and a whole lot of economic monodependence.... The government does not understand what the issue is about. It is first of all about a New Culture of Water and the outright rejection of the PHN.... [And s]econd: Catalonia needs an energy model, it does not have one; an energy model that foments renewables, implementing them in a manner that respects natural patrimony. This model should transcend the models of the 1960s and 1970s. It should advocate for small and medium units of generation, located in the vicinity of those areas of higher consumption and demand.[27]

Indeed, the southern platforms were not simply rejecting specific projects; rather, they were arguing for a new approach to water policies—a New Culture of Water—and a new energy model—a New Culture of Energy.

A New Culture of Water

The doctrinal core of the proposed PHN was an old thesis of the *regeneracionistas*, the enlightened elites of the turn of the twentieth century (see chapter 2), for whom Spain was divided into two kinds of regions: those with water-deficit—"dry Spain"—and those with water-excess—"wet Spain." For the *regeneracionistas*, fixing this imbalance was the key to Spain's national development. Although never realized, this project remained alive in the minds of Spain's ruling classes throughout the twentieth century. In the 1990s, successive center-left and right wing governments fought for the approval of the PHN, a plan that was to put a definitive end to Spain's hydraulic disequilibrium. The PHN was nothing short of gargantuan in scale and scope. It involved complex, enormous new hydraulic infrastructure connecting river basins on a scale so large that the government acknowledged from the start that the plan could only be realized if the EU co-funded it. The central, and most controversial, project within the PHN was the transfer of water from the River Ebro to the conurbation of Barcelona and the Spanish Levant (regions of Valencia and Murcia).

The provisional approval of the PHN unleashed conflicts between donor and recipient regions that came to be known as "water wars." In regions situated within the Ebro's river basin, primarily Aragon and Southern Catalonia, the PHN was received with outrage, unleashing large mobilizations calling for the protection of the river and the withdrawal of the project. These complaints elicited an intransigent, authoritarian response from the government, epitomized by the Spanish Minister of Public Works, who said that the approval of the PHN would be a "military march," only to add later that it would be approved *por cojones*, a crude testicular expression.[28]

Moreover, the whole project was surrounded by inflamed nationalist rhetoric, with recurrent statements such as a newspaper headline quoting then Prime Minister José María Aznar: "The PHN is not a partisan project, but one that is vital for the cohesion of Spain."[29] For their part, the governments of the regions of Murcia and Valencia, in the hands of Aznar's PP, relentlessly campaigned for the PHN, accusing the donor regions of lacking in national solidarity.[30] This campaign reached its zenith on March 2, 2002, when the Valencian government organized and funded a demonstration under the motto *Agua para todos* ("Water for all"). Tellingly, the demonstration was a failure in terms of attendance. In Catalonia—the only region that was to be both donor and recipient of water flows—public opinion was overwhelmingly against the PHN.

Despite the underlying logic of the project and the rhetoric about national unity and interregional solidarity, the fact was that the donor regions were hardly wet and wealthy. The supporters of the PHN were demanding solidarity from regions that were generally poor, dry, peripheral, and already contributing more than their fair share to the energetic and material metabolism of the country. The main virtue of the anti-transfer movement in Aragon and Southern Catalonia was pushing the debate beyond the logic of regional competition.[31]

The Platform for the Defense of the Ebro (PDE) argued that water scarcity was not an absolute physical fact. Rather, it was a socioeconomic issue rooted in a wasteful and authoritarian model of water management, embodied by the PHN and founded on the domination of nature and society. In place of this model, the PDE called for a "New Culture of Water," an expression that identified a new socioenvironmental paradigm based on the principle of sustainability. In the words of one of its most notable advocates, economist Pedro Arrojo, this new culture should "recognize and value the ecological foundations and services generated by rivers, lakes, aquifers and wetlands, as well as the sociocultural, emotional, and identity values linked to that natural patrimony, all within an ethical framework governed by the principles of sustainability and equity."[32] Whereas the PHN was premised on the management of water as a generic, bulk resource, the New Culture of Water advocated for an ecosystems approach based on conservation and efficiency: the problem was not one of scarce supply, but rather of excessive demand. Perhaps as important, the New Culture of Water called for the democratization of water politics, in opposition to the technocratic authoritarianism that had historically dominated Spain's hydraulic policies.[33] In advocating for the valorization of water ecosystems and the principle that water should be managed within, rather than across, river basins, the New Culture of Water was consistent with a model of endogenous development and reminiscent of the idea of *autogestió*.

Perhaps the principal merit of the PDE and its allies was how they connected their demand for a new approach to water policy with a denunciation of the economic interests that lay at the heart of the PHN. Against the government's argument, which depicted Southern Catalans as selfish citizens depriving their fellow countrymen of water, the PDE took pains to emphasize that the water sent to Barcelona and the Levant would not serve urban dwellers and small farmers, but instead would benefit a select few in the thriving real estate and tourist sectors, as well as a non-sustainable agroindustry requiring cheap water and cheap migrant labor.[34] Beyond these sectors, the main beneficiary of the PHN was to be the "brotherhood of cement," what Antonio Estevan called "the two big economic agents that have traditionally controlled hydraulic politics in Spain: big construction companies and the electricity corporations."[35] In unveiling the corporate interests that were to be the main beneficiaries of a project that had been presented as politically neutral and good for the general progress of the country, the PDE overcame a rigid understanding of the city-country division. Arguing that the PHN reinforced this division and peripheralized Southern Catalonia, the PDE insisted that its antagonists were not urban dwellers, but the economic and political interests whose power and profits were based on a city-country divide.

A New Culture of Energy

From a narrow environmental viewpoint, the contrast between wind farms and other existing, nonrenewable energy facilities in Southern Catalonia—natural gas power stations, nuclear plants, and fossil fuel pipelines—could not be starker. However, the platforms of Terra Alta and Priorat gave voice to the experience of many Southern Catalans, for whom the arrival of wind energy was not lived as a rupture but as a continuation of old patterns. By doing this, the southern platforms invite us to interrogate the kind of energy transition involved in the development of wind energy in Spain. This crucial insight was only clearly formulated *after* the platforms entered into conversation with other groups, especially the PDE, in the context of the Southern Revolt.

Indeed, the fall of 2000 was a crucial point of inflection in the battle over wind energy development in Southern Catalonia. The emergence of a broad mobilization reinforced the visibility of the wind conflict and supported the claim of the wind platforms.

"In the morning the government told us that we had to save the universe by producing renewable energy, and I am an environmentalist," Núria explained. "But then, in the afternoon it was blessing the biggest thermal power plant in Europe. They made it very easy for us!"

The platforms of Priorat and Terra Alta had always argued that climate change fears were not the primary driver of wind energy development—instead, they asserted, the development of wind energy was predicated on continued exploitation of a peripheral area targeted for the extraction of profits because it already was an electricity-producing hub. The plan to build a natural gas power plant just a few kilometers downstream from Ascó's nuclear reactors reinforced that argument. So did the scale of the project: with an installed capacity of sixteen hundred MWh, the Móra plant was to produce far more electricity than all the wind farms projected for 2010 in the Wind Map for all of Catalonia. Later events would lend credence to the platforms' arguments: in 2001 a large Spanish electric utility announced plans to build two more state-of-the-art natural gas plants in the region; to make matters worse, a few months later the Catalan government released a project to install a large-scale industrial waste dump in Ribera, the county squeezed between Priorat and Terra Alta, where Ascó is located.

The Southern Revolt also helped the platforms gain a broader understanding of their struggle and refine their strategy. Certainly, the platforms of Priorat and Terra Alta had always emphasized issues of scale and planning—wind farm over-crowding, regional equilibrium, and ecological balance—so it would be unfair to reduce them to selfish, "not-in-my-backyard" organizations. Yet it is also true that, up until 2000, their actions were basically defensive and their main message was "Not here!"[36]

The eruption of the Southern Revolt, their dialogues with other platforms, and the struggles they participated in around the PHN pushed the wind plat-forms to adapt and extend their strategies. It became clear that it was not enough to say that they did not want wind power plants, they had to offer a countervail-ing vision.

The platforms called this vision a "New Culture of Energy." Its most com-plete formulation came out of a meeting held in Tortosa in November 2003 and attended by representatives of platforms—Priorat, Terra Alta, and Ribera—and left-wing political parties. These stakeholders signed and released a document with the convoluted title "Manifesto-Agreement for a New Culture of Electric Energy."

The first part of the document was an indictment of the frameworks and policies governing the electricity sector, the patterns of electricity production, and the development of wind energy; the second section proposed an alternative. More specifically, the first part of the document observes that: (a) the liberaliza-tion of the electricity sector has favored the interests of big utilities; (b) energy planning uncritically assumes a growing electricity demand, instead of foment-ing energy conservation and efficiency; (c) energy policies are doing nothing to diminish the reliance on nuclear power and fossil fuels, for the government shows no intention to phase out nuclear plants and is encouraging the construction of new natural gas power plants; (d) the distance between the areas where electricity is consumed and the areas where electricity is generated reproduces a centralized electricity system, plagued with inefficiencies, that perpetuates territorial tensions; (e) Southern Catalonia concentrates not only most of the existing electric generation, but also most of the proposed new production; and (f) wind energy development is taking place within a centralized energy system that encourages the creation of large wind power plants that endanger spaces of high natural and cultural value.

The second part of the document is a collection of demands. Taken altogether, these constitute what the platforms dubbed a New Culture of Energy: (a) crafting and passage of a new Catalan Energy Plan based on the principles of conservation and efficiency; (b) dismantling of nuclear and coal power plants; (c) higher taxes to nuclear and fossil fuels; (d) creation of a new energy system based, on the one hand, on developing renewables and, on the other hand, on decentralization, reducing the distance between production and consumption; (e) respect for spaces of a high natural, landscape, or cultural value and for the principle of territorial equity, demanding a fair redistribution of ecological loads and benefits; and (f) implementation of participation-based mechanisms in the making of energy policies. Finally, the document calls for a wholesale moratorium on the planning and installation of *any* electricity production unit until the overarching energy plan has been approved.

The influence of antinuclear and environmentalist thought on this New Culture of Energy is evident. Indeed, its decentralized model rehearses the arguments and vision of the antinuclear activists of the 1970s and 1980s that we saw in chapter 2. But by linking the peripheral situation of Southern Catalonia to the existing policies of water and energy management, the southern platforms were also uncovering the relationship between these policies and the dominant patterns of accumulation. They showed that the PHN, the Wind Map, and Enron's power plant only made sense within a certain structure of accumulation that was, at the same time, an ecological regime.

The Southern Revolt in Perspective: Spain's Second Miracle

The Southern Revolt was not an isolated phenomenon, but the forefront of a truly large number of local territorial and environmental conflicts that mushroomed in Catalonia during the first decade of the twenty-first century. During this period, one could not visit any of the forty-one Catalan counties without encountering at least one platform opposing a specific infrastructural or construction project: airports, roads, waste dumps, golf courses, seaside resorts, and urban renewal schemes were all among the targets.[37] This proliferation of conflicts received intense attention from Catalan social scientists, which connected the mobilizations as a single, if largely uncoordinated movement.[38] Local activists, though their platforms tended to operate independently, shared this vision of heterogeneous unity, identifying their struggles as part and parcel of a nation-wide "movement in defense of the territory."

The movement has a complex genealogy, connected with the traditions of peasant activism, environmentalism, and urban activism, as well as the antiglo-balization movement and Catalan separatism. At the ideological level, we find a similar broadness, allowing the "defense of the territory" to hold multiple mean-ings: the valuing of the local against Barcelona's dominance, a centralized state, and globalizing trends; the protection of the environment against capitalist deg-radation and modernist instrumentality; or the defense of Catalan land against foreign economic and political structures. Perhaps the two most basic character-istics of the movement were its unrelenting criticism of the dominant dynamics of economic development, often anchored in clear anti- or de-growth positions, and its demand for participatory forms of democracy, coupled with an emphasis on autonomous forms of organization and *autogestió*, reflected in the platforms' open, horizontalist structure.

The Southern Catalan ecological distribution conflicts held a unique posi-tion within this wider mosaic. Nowhere else did these conflicts give rise to the Southern Revolt's brand of large, popular movement. This movement inspired other struggles, which modeled their actions and goals on what was done in Southern Catalonia. In a paradox, the Southern Revolt's centrality stemmed from the region's peripherality; its marginality provided a privileged perspective for understanding, and therefore critiquing, the structures and dynamics of political economic power in Catalonia and in Spain.

The so-called "movement in defense of the territory" emerged as a response to the patterns of development of what I call the Second Miracle, a period of high rates of economic growth roughly spanning between 1995 and 2008. Although the movement has had a strong, enduring influence on Catalan politics, the pro-liferation of territorial conflicts has dramatically slowed down in recent years, coinciding with the deceleration of the Spanish economy since 2008. The move-ment may in fact be read, using Polanyian language, as a countermovement, an unplanned social defense against a structure of accumulation based on a frenetic use and abuse of space and nature.[39] The housing bubble, the ecological degra-dation of the seaside, the dramatic underuse of new, large-scale infrastructures such as airports, power plants, and high-speed trains—as well as the corruption that accompanied these projects—are all part of the present-day consequences, a damning testament to the Second Miracle's effects on Spain's environment. In the next chapter, I will deal at length with the Second Miracle, exploring its per-vasive influence on wind energy development and the entire Spanish electricity system. For now, I will limit myself to a single argument: the Second Spanish

Miracle was a deeply ecological process. This is, I believe, the ultimate lesson that we can learn from the Southern Revolt.

Habits of language make my task difficult. We are used to describing the *effects* of "the economy"—or "technology" or "society"—on "the environment," from habitat degradation to climate change. Yet this presupposes two distinct, independent entities acting on each other. Such way of thinking stands in the way of understanding capitalism as a socioenvironmental system. Jason Moore's notion of *ecological regime* allows us to escape dichotomous thinking and advance toward a dialectical understanding of capitalism as a way of organizing nature. Ecological regimes are: "Those relatively durable patterns of governance, technological innovations, class structures, and organizational forms that have sustained and propelled successive phases of world accumulation.... At a minimum, these regimes comprise those markets, productive and institutional mechanisms necessary to ensure adequate flows of cheap energy, food, raw material, and labor-power to the organizing centers of world accumulation.... The town-country antagonism—overlapping with but distinct from the core/periphery divide—is the pivotal geographical relation. In this, ecological regimes signify the historically stabilized process and conditions of extended accumulation."[40]

The notion of ecological regime, then, refers to durable sets of relations organizing the metabolism—that is to say, the material exchanges between humans and the environment—of any given political-economic order.[41] These relationships are not contingent but a central constitutive element of that order. Whereas Moore thinks about ecological regimes at the scale of the world system, my aim is to apply the concept to a specific national economy.

Thus, in chapter 1, I argued that the *desarrollismo* of the 1960s, Spain's First Miracle, inaugurated a new ecological regime in Spain. Massive rural exodus and proletarianization, the birth of mass tourism, the frantic construction of dams, the early development of agroindustry, and the increased importation of fossil fuels were all foundational processes of this new ecological regime. As I described in that chapter, these processes index a rearticulation of the town/country divide that spurred exponential growth in certain urban-industrial poles and produced rural areas, such as Southern Catalonia, with no apparent economic function or value. As described in chapter 2, this resulted in a sharp geographical division between centers (big cities and coastal areas) and peripheries (the rest). It also signaled the commanding position of banks, construction companies, electricity utilities, and, to a lesser extent, the tourist and some heavy industry sectors— steel, cement, chemicals—in the modern Spanish economy.

The Second Miracle was not founded on the creation of a new ecological regime, but rather on the intensification of the existing regime that stretched to new limits. This was part of a process of consolidation and partial restoration of class power and state centralization. Indeed, the Second Miracle may be seen as a reassertion of the power of a handful of economic sectors, following a couple of decades—roughly from 1975 to 1995—of relative stagnation and corporate restraint. This process relied on extending the Spanish ecological regime in four directions: massive inflows of European surplus capital, immigration from countries in the Global South, the corporate takeover of strategic sectors in many Latin American countries, and what David Harvey calls a new "spatial fix," extending the internal geographical frontiers of accumulation.[42] The latter was the key emblematic process of the Second Miracle. Properly speaking, it did not suppose an expansion of value and productivity. It consisted of an expansion of the frontier of *ground rent valorization*, leading to asset inflation and concurrent mechanisms of monopoly rent extraction.

The functioning of this frontier mirrored the structural position of the Spanish elites and the rentier mechanisms of accumulation that sustained their position. All the big winners of the Second Miracle—banks, electric utilities, construction companies, and the tourist and real estate sectors—engaged in a construction binge, funded by cheap credit and focused on residential units and large-scale infrastructure. This intense mobilization of space led to unprecedented inflation in real estate prices and the escalation of private and corporate indebtedness, but also fed a demanding metabolism that gobbled energy and materials. The geographic pattern of construction was both an (intensified) continuation and an extension of the spatial patterns of the First Miracle, privileging central areas (urban and coastal), and pushing into peri-urban areas and almost the entirety of the Mediterranean coast. New areas of ground rent valorization sustained a form of rent extraction crucial to the reproduction of the dominant structures of accumulation: the "rest" of the country was *iterated* as a periphery.

The Second Miracle was in this way characterized by an uneven production of space and nature that reasserted the division between central places, subject to ground rent valorization, and peripheral spaces, subject to devalorization and relegation to wastelands performing functions of extraction and sink. Yet the Second Miracle's increasing material demands and its extension of ground rent valorization also made peripheries relatively scarce.[43] It thus redoubled the pressure on those peripheries, as the PHN exemplifies, in order to sustain the existing patterns of accumulation. By measuring the exchanges of materials and energy between Spanish regions during the Second Miracle, Óscar Carpintero has been able to provide quantifiable evidence of the miracle's production of ecological and spatial unevenness: "From the perspective of the physical *extraction* of energy and materials, there is a sharp hierarchical growing contrast between regions, with increased concentration of population and economic activity in regions economically central such as Madrid, Catalonia, Basque Country and Valencia. Those regions that concentrate the bulk of physical extraction, however, do not at all coincide with the central regions (except for Catalonia), but rather with those economically peripheral.... There are two kinds of regions in Spain: those specialized as tap and sink, and those accumulating and consuming."[44]

The value of these measurements is that they show how apparently balanced economic exchanges actually embody huge ecological imbalances. This means, from a metabolic perspective, that capital undervalues material and energy flows and peripheries remain in the penumbra of capital's law of value. The Catalan exception that Carpintero documents is, on the other hand, easy to pin down: its metabolic balance is based on an internal imbalance between the Catalan core and its periphery. The imbalance in material flows between core and peripheral Spanish (and Catalan) regions is also intuitively easy to predict: available water, food, and inorganic materials are extracted from peripheries, where there is little demand for them, and sold to the centers where population and economic activity are concentrated. The same logic applies to electricity, yet we should note that in this case the pattern cannot be explained in terms of resource abundance: with the exception of hydroelectricity, Spanish peripheral regions do not possess the primary resources—uranium, fossil fuels—to generate electricity. Renewables are different, in that peripheral regions do possess wind and sun, but so do core regions, so that does not explain much, either.

The solution to this riddle goes to the heart of the core-periphery dynamic articulating Spain's ecological regime: peripheral regions offer plenty of undervalued space. Conversely, they contain a small number of undervalued people. We may remember that one of the two reasons explaining the location of the nuclear reactors in Ascó was the availability of water. Yet this explanation, on its own, is insufficient: Spain may not have a lot of fresh water, but it does have a lot of seawater. The problem with seawater is that the places where it is located are too valuable, prime real estate. Ascó received that hazardous facility precisely because it was not a valuable place. The same dynamic is at play with wind energy: it requires large amounts of cheap (i.e., devalorized) space, and this became increasingly scarce during the Second Miracle.[45]

One would assume that wind farms are located in areas with an abundance of wind, and hence that wind abundance is the best predictor of their location. This correlation tends to hold at the intraregional level, but if we look at the whole of Spain, it is readily apparent that low real estate value—that is, peripherality—is a far better predictor of wind farm location than is wind abundance. Indeed, five out of the seventeen *comunidades autónomas*—the two Castilles, Andalucia, Galicia, and Aragon—concentrate 78 percent of wind energy production. In contrast, financial and industrial powerhouses, such as Madrid and the Basque Country, or tourist hubs, such as the Balearic and Canary Islands, all produce way less than 1 percent.[46] This is an almost perfect reversal of the geography of accumulation, which is concentrated in urban, tourist, and coastal areas. Again, with a total installed wind energy capacity (5.5%) roughly equivalent to its surface area (6.3%), Catalonia is somewhat of an exception.[47] And again, the explanation for this anomaly is the existence of an internal periphery. There are no wind farms anywhere near the Catalan coastline or in Empordà, the windiest but one of the wealthiest Catalan counties. During the Second Miracle, cheap space became increasingly scarce in Catalonia, not only because of the extension of the frontier of value, but also due to mobilizations such as the Southern Revolt, opposing infrastructural developments while resignifying their territory as places of dignity and value.

The Southern Revolt was a protest against the iteration of Southern Catalonia as a peripheral space. The fact that it peaked during the middle years of the Second Miracle shows that this "miracle" was built on an intense appropriation of space and resources. While housing construction, skyrocketing real estate prices, and the inauguration of large-scale emblematic projects dominated the headlines, the Southern Revolt tells us that those flashy achievements required the devalorization of places such as Southern Catalonia. This view emerged from practical knowledge, a practical consciousness grounded in an experience of marginalization and peripheralization. Stressing the connection between its struggle and the feats of the Second Miracle, the Southern Revolt told us that the financialization of the Spanish economy, the growth of the tourist sector, and the looming housing bubble were all ecological processes, as much as the extension of renewable energy production or the transfer of water to core areas. The Southern Revolt told us that Southern Catalonia was a peripheral region *in* the Second Spanish Miracle, but in no way peripheral *to* it. Perhaps most crucially, the Southern Revolt told us that peripheral regions can provide a privileged perspective for understanding Spanish capitalism.

Priorat and Terra Alta: Divergent Paths

In June 2002, the Catalan government passed a second Wind Map.[48] This revised Wind Map introduced few novelties. These were limited to strengthening the obligation to perform environmental impact assessments and to eliminating the possibility of installing wind farms in the most contentious locations, such as Montsant and Pàndols-Cavalls. Yet like the first one, this second Wind Map targeted Sothern Catalonia as the main area for wind energy development, considering most of its territory as compatible with wind farms. Unsurprisingly, both platforms rejected this new regulation. But by then the two platforms and the counties they represented had already taken opposite paths.

The struggle against wind energy development had opposite outcomes in Priorat and Terra Alta. In Priorat, it strengthened the hegemony of the project of endogenous development; in Terra Alta, it perpetuated the division between the two sides of the conflict. The inhabitants of Terra Alta tend to explain this divergence as a reflection of the extreme fractiousness of Terra Alta's political landscape, which translated into antagonistic attitudes toward the project of "war tourism" and heightened distrust between municipalities. As one popular saying in the county claims, "Terra Alta: twelve villages, twelve tribes." Nonetheless, the most important factors accounting for these divergent paths were the unambiguous support of the county's wine sector for the Platform of Priorat and the platform's organizational skills. These skills were evidenced in the construction of a solid, encompassing discourse around the notion of landscape, and in the creation of a broad consensus around a credible scenario for wind energy development in the county.

In 2001, the platform brokered a Wind Agreement supported by all political parties and the DOs. The agreement was simple: it limited the number of wind farms that could be installed in the county to two and asserted that their optimal location was in the northwestern corner of the county, where wine production is least prominent. In Terra Alta there were also talks of establishing a similar sort of agreement, but lack of support and internal divisions—many in the platform were not willing to accept any wind farm—prevented the platform from even presenting a proposal.

On the other hand, the Platform of Priorat found, in the notion of *landscape*, a category that could overcome some of the contradictions we have explored so far. Landscape, for the group, included "natural," "cultural," and "identity" values, bundled into a unique configuration of elements built over time, thus harmonizing the equivocation between the two incommensurable yet partially connected languages of valuation that we saw earlier. Indeed, this notion of landscape overrode the contrast between plains and mountains, while making the ecocentric vision of the naturalist groups compatible with the peasant environment of the local population, given that the terraced vineyards were taken as the defining idiosyncratic feature of Priorat's landscape. Needless to say, the celebration of the landscape found its most avid advocates among winemakers, eager to link their product with the *terroir*. The Platform of Priorat successfully presented the territory as a source of dignity and self-esteem, and as an economic resource. Roser Vernet, a leading voice in the platform, argues: "[Through our struggle] our landscape has gone from being considered a symbol of poverty to be taken as an element of identification and cohesion, a patrimony to be recovered and valued. In this process, a primordial role has been played by those economic sectors strongly based on the endogenous resources [*recursos propis*] of the county, principally the wine sector.... We finally see a possibility for moving forward ... and our landscape becomes inseparable from this development model.... It becomes a value in itself."[49]

Indeed, soon after 2001, Priorat entered into a new phase in its history, and this new present is largely free of wind farms. Although Priorat's Wind Agreement does not have juridical force, every time that a new, large-scale wind energy project has been announced for Priorat, the platform has been able to stop it by mobilizing the social and political support gathered around the agreement. In fact, in recent years the Platform of Priorat has, for the most part, reoriented its action toward the recognition of the value of the county's landscape, an endeavor that is currently being channeled in the campaign to get UNESCO to designate Priorat as a "Mediterranean mountainous agrarian cultural landscape."[50]

Earlier, I have argued that the ecological regime of the Second Miracle was based on distinguishing between two kinds of spaces: those subjected to valorization, and those condemned to devaluation. Priorat has been able to enter the first group, converted into a *terroir* through a process of self-valorization. The new situation of Priorat is not without conflicts and challenges—growing land speculation, new processes of class differentiation, continuing difficulties for generational turnover in peasant households—yet the pros outweigh the cons, and for the first time since the war, the county has been experiencing a demographic recovery.

Yet, if Priorat won, Terra Alta lost. Since 2004, the struggle against wind energy development in Southern Catalonia has essentially been reduced to this county, where as many as eight large wind farms have been put into operation in the last few years, with several more on the way. With the actual construction of wind farms, the center of the conflict over wind energy development has shifted. The institutional arena, dominated by organized forms of protest, has lost centrality in favor of smaller scale, more subterranean conflicts within villages and between developers, landowners, and municipal governments. The fact that most of the wind farms have been installed in agricultural land has just added to the intestine character of these conflicts, as we will see in successive chapters. The general context in the region has also changed. Up until 2004, the wind conflicts in Terra Alta were part of a broader struggle, first shared with Priorat, later with the entire region and, to an extent, the whole country. Yet by the time that the wind farms were built in Terra Alta, this broader horizon had largely disappeared: Priorat had fixed its problem, the Southern Revolt had lost steam, and after 2008, even the miracle seemed a distant memory. The inhabitants of Terra Alta felt alone, abandoned to be a wasteland.

Yet the Struggle Keeps Going

After its approval in 2002, the PHN was delayed for two years, mostly because the EU was reticent to fund the project. This allowed a new social-democratic Prime Minister, José Luis Rodríguez Zapatero, to withdraw the project soon after his election in 2004. This victory for the Southern Revolt, combined with the election of new, left-wing governments both in Spain and Catalonia, help explain why the Revolt lost steam after 2004. It did not die, though; government and private corporations kept targeting the region for the installation of contentious infrastructures, triggering widespread opposition. In addition to new wind farms, electric lines and conventional waste dumps, two of these infrastructures stand out: a submarine natural gas storage facility—the largest in the country—situated in front of the Ebro Delta and a state-owned nuclear waste dump in Ascó.[51]

The fate of these two projects is telling. The government approved the natural gas storage deposit. But once its execution was well advanced, the government determined that the facility was unsafe, compensating the developer—incidentally, the largest Spanish construction company—with a sum of money so astronomical that it had to be revised by the European Union.[52] As for the nuclear waste dump, popularly known as the "nuclear cemetery," the government ended up moving its location to a different Spanish region, fundamentally motivated by intense opposition from Southern Catalonia. This mobilization, which took place during my fieldwork, demonstrated the depth of the Southern Catalan antinuclear tradition and the vitality of the Southern Revolt.

One of the central events of the campaign against the nuclear cemetery took place in the fall of 2010, when a crowd gathered from all over Southern Catalonia marched through the streets of Ascó. Núria gave one of the speeches that closed the demonstration, powerfully capturing so many Southern Catalans' indefatigable quest for dignity:

> We're here again, although we could and would like to be elsewhere, building our territory, and a country [Catalonia] that we also want to be ours ... [But] once again, they don't let us use our energies—and these ones really are renewable—to create and build. Because we have to waste them, day after day, defending the dignity of our land and, no one should forget it, the dignity of a whole country.... We are losing the possibility of breaking away from the fatalism that comes from seeing how, today as much as yesterday ... undemocratic forces [*poders fàctics*], from energy to finances, step on the so-called public democratic powers and impose their will, based on the profit for a few and the risk for many—and the spare change for those who shut up and acquiesce. This has been going on in our counties for too long. Here we keep accumulating grievances that generate, on the one hand, passivity and servility, and on the other hand, disenchantment and outrage.... Our demand, supported by a majority, has been loud and clear: we don't want the nuclear cemetery. And we also demand respect for our despised landscapes, for our abused rivers, for our threatened peasantry, in sum, respect for our people.... We will not give up, because we know that, together, we can stop anything we want. Out of our outrage we'll build renewed hope. *Visca la terra!*

Notes

1. Saladié (2011a).
2. See, for instance, Escobar (2006), Martínez Alier (2009), and Robbins (2012); for the specific case of Southern Catalan wind, see Zografos and Martínez Alier (2009).
3. Tomàs (2002: 9).

4. Narotzky (2016a: 82).

5. See, for instance, "A vueltas con el mapa eólico catalán" in *Energías renovables*, issue 9, 2002.

6. On abstract social nature, see Moore (2015: 193-217); see also Ferry and Limbert (2008) and Harvey (2001).

7. *La Serena*, April 1, 1997, p. 3.

8. Saladié (2011a).

9. Lefebvre (1991).

10. Altvater (1993: 69). This nonvalorizable part of the socioenvironmental world roughly corresponds with what James O'Connor (1998) calls "conditions of production."

11. Hornborg (2011: 27–44).

12. On languages of valuation, see Martínez Alier (2009).

13. On the interpenetration between the struggle to preserve the memory of the Civil War and the wind energy conflict, see Castell and Nel·lo (2003).

14. Appadurai (1986).

15. Benjamin (1968b).

16. For an analysis of the relationship between inalienability and marketability, see Franquesa (2013b).

17. Saladié (2011b).

18. The other two will briefly appear in chapter 5: the first one built in 1999 by a developer whom I call Gerard; the second, a wind farm built in Tortosa in 2004–2005 by a developer whom I call Alfredo.

19. See, for instance, the long article "La eclosión de los parques eólicos divide al movimiento ecologista en España," *La Vanguardia*, November 14, 1999. See also Sempere (2008).

20. For two different perspectives on this process of separation, see Descola (2013: 57–88) and Smith (2008).

21. Ingold (2000); see also Latour (2017: 111-145).

22. Ingold (2000: 218).

23. Strathern (2004).

24. De la Cadena (2015).

25. Southern Catalonia's position within the Catalan body politic has historically been rather eccentric and subject to debate. For an analysis of how the Southern Revolt shifted this debate, see Guiu (2008).

26. CiU voted against the PHN, triggering the PP to retaliate by voting against the Wind Plan and the Enron project, actually leading to the defeat of all three measures. The two parties had been allies in the Catalan and Spanish governments since 1996.

27. *Diari de sessions del Parlament de Catalunya*, March 8, 2001, page 35.

28. On the authoritarianism of the PHN, see Aguilera Klink (2003).

29. *El País*, February 4, 2001.

30. On these campaigns, see Almenar and Bono (2005).

31. On the Southern Catalan PDE, see Pont (2002) and Alfama and Miró (2005).

32. Arrojo (2006: 163).

33. Aguilera Klink (2003).

34. Del Moral Ituarte (2003).

35. Estevan (2003: 48). On the interests of banks, electric utilities, and construction companies in the PHN, see De Prada (2003).

36. I borrow the expression from Nel·lo (2003).

37. For an exhaustive compilation of these territorial conflicts, see the *Anuari Territorial de Catalunya*, accessed October 3, 2017, http://territori.scot.cat/cat/anuari.php.

38. Among others: Garcia (2003), Nel·lo (2003), and Alfama et al. (2007).

39. Polanyi (2001).

40. Moore (2015: 158).

41. On metabolism, see Fischer-Kowalski (1998) and Foster (2000).

42. Harvey (1999).

43. This is made evident in the concurrent increase of agrarian land prices and loss of farmed land in several Spanish regions (although not in Catalonia) between 2000 and 2009 (Soler and Fernández 2015: 22–32 and 121–131).

44. Carpintero et al. (2015: 2–3).

45. For the sake of clarity, I use devalorization (or disvalorization) and devaluation as equivalent terms, using them to indicate the *relative* loss of value embodied in material forms without destruction of those forms (for a definition see Harvey 1999: 84). For a finer distinction between the two terms, see Cleaver (1992).

46. AEE (2016: 38).

47. This exception is, however, only relative, for Catalonia possesses 16 percent of Spain's population. Valencia shows an analogous pattern.

48. Its official name was "Decret regulador de la implantació de l'energia eòlica a Catalunya."

49. Vernet (2008: 98–99).

50. See the candidacy's website: www.prioritat.org

51. In 2016, the government announced a third project that will test the resisting capacity of Southern Catalans: a new version of the PHN. As I write, the PDE has already organized massive demonstrations against the project.

52. "Bruselas solicita información a España por la indemnización de Castor," *El País*, January 21, 2016.

5 Wind Bubble

Cerç de matinada, trons de vesprada [Wind in the morning, storms in the evening].
—Southern Catalan popular saying.

IN THIS CHAPTER, we depart from the territorial conflicts of Southern Catalonia and enter the offices and social milieu of wind developers and turbine manufacturers in order to understand how a renewable energy source can take on the characteristics of an extractive one. In 2013, wind became the main source of electricity in Spain, an unprecedented feat in modern history. Yet in that same year, after fifteen years of sustained growth, and largely as a result of the elimination of production subsidies to renewables, wind energy development came to an abrupt halt; barely any new capacity has been installed since then and there are no signs that the situation may reverse in the short to medium term. This pattern of behavior may seem puzzling: boom-and-bust cycles have historically been associated with extractive economies, from guano to shale gas, whereas wind energy represents the renewable form of energy production par excellence.[1] Yet the solution to this puzzle does not rest on the technology or the energy source, but rather on the political economy of wind energy production.

The main argument of this chapter is that the precipitous rise and fall of wind energy—what I call the *wind bubble*—should be understood as a byproduct of the Second Miracle. Adding to the ecological analysis of the Second Miracle that I offered in chapter 4, we will see that wind energy became progressively inserted into the main dynamics of the miracle—rent-seeking behaviors, the oligopolistic position of large utilities, reliance on cheap credit, and pervasive political clientelism—and these dynamics conferred wind energy development an extractive character. A more general point stems as a corollary of the analysis: the key issue for the understanding of renewable energy does not lie in the intrinsic character of its form but rather in the way it is organized and governed (for instance, as more or less centralized, or more or less democratic). Thus, the chapter contributes to one of the main arguments of the book: energy transitions are not technological shifts, they are sociopolitical processes fraught with conflict.

The Promise of Renewables

To explain how wind took on an extractive character in Spain, I must first begin with a discussion of the imaginary of renewables as a decentralized and decentralizing technology, capable of democratizing both the energy system and society at large. Debates around renewable energy have historically revolved around its promissory capacity to transform the existing energy system, breaking with a legacy of noxious emissions, large-scale infrastructure, and centralized mechanisms of decision making. Renewables have held the promise not only of a greener and more decentralized future, but also of an increasingly democratic society. This *ruptural vision* of renewable energy is, however, based on a fair amount of technological optimism and energy fetishism. Rather than taking the technical features of renewables as a point of departure, our analysis must first focus on the social structures and relations that control the energy system. Renewables are a field of struggle: they have disruptive potential, but they can also reproduce the characteristics of the inherited energy system. In this section I focus on three moments in the long debate over renewables: the early twentieth-century dreams about the potential of hydroelectricity, the 1970s debates around nuclear and renewable energy, and contemporary discussions on the shape and direction that wind energy should take.

Hydroelectricity and Technological Optimism

During the first decades of the twentieth century, hydroelectricity was the object of utopian dreams. Technologically optimistic authors such as Petr Kropotkin, Ebenezer Howard, and Lewis Mumford hailed hydroelectricity as crucial for leaving behind the insalubrity, degeneration, pollution, and ugliness of a first industrial revolution reliant on coal, steam, and mining, a period that Patrick Geddes defined as the "paleotechnic" phase of human development.[2] These utopian, environmentally minded intellectuals—all of them with some level of sympathy toward anarchist ideas—thought that hydroelectricity was going to make possible, even inevitable, social redemption. Howard's garden-cities, Mumford's neotechnic civilization, and Kroptkin's decentralized productive system—where hand and brain, as well as industry and agriculture, would be reintegrated—all would be hastened by this technology that liberated human activity from the locational servitudes of a productive system revolving around coal and iron. With hydroelectricity, gray skies and overcrowding would give place to clean and smart production; economic growth would be decoupled from the depletion of energy stocks; activities and population would be freed to distribute across the territory.

In his 1934 *Technics and Civilization*, Lewis Mumford made an especially compelling case for this vision. He believed that coal was the origin of the centralization and inhuman scale of industry—"the steam engine tended toward monopoly and concentration; [but] wind and water were free"—engendering an extractive mentality that extended all throughout the whole of society: "The animus of mining affected the entire economic and social organism.... The reckless, get-rich-quick, devil-take-the-hindmost attitude of the mining rushes spread everywhere: the bonanza farms of the Middle East in the US were exploited as if they were mines, and the forests were gutted out and mined in the same fashion as the minerals that lay in their hills. Mankind behaved like a drunken heir on a spree."[3]

This extractive logic, Mumford adds, also applied to the relationships between humans: "Human beings were treated with the same spirit of brutality as the landscape: labor was a resource to be exploited, to be mined, to be exhausted, and finally to be discarded."[4] Ultimately, for these three authors, the extension of electricity held out the promise that society could not only overcome the greedy mentality of the industrial age, but also usher in a new period of fraternal relations with the environment and among humans. Their optimistic determinism would soon prove ill founded.

Indeed, as we saw in chapter 2 in the Spanish case, during the first decades of the twentieth century, electricity systems were progressively centralized and concentrated. Writing at the end of this process, Mumford realized that all that electric possibility had been coopted by the rationality of a former technological regime: "The damage to form and civilization through the prevalence of these habits of disorderly exploitation and wasteful expenditure remained, whether or not the source of energy itself disappeared."[5] Hydroelectricity and long-distance transmission did open new possibilities, as Kropotkin, Howard, and Mumford clearly saw, but they also allowed for the reworking and extension of existing patterns of accumulation and unequal power relationships. In short, any examination of technology remained abstract and narrow-minded unless it was coupled with the analysis of social relations of production.

Energy Paths and Resource Determinism

In the 1970s, oil crises created the context for a societal debate about the future of energy systems. There were two options on the table: nuclear energy and renewables. Nuclear power won that battle. That victory had much to do with the ease with which this form of energy could be inserted into the rationality governing the energy system, as Debeir *et al.* argued in the 1980s: "The efficiency of the dominant energy system—whose logic consists of thinking of energy demand as having a homogeneous and primary character rather than a heterogeneous character corresponding to real consumption—remains low. Nuclear power is in no way the instrument of an alternative energy future, it is simply the replacement of pre-1973 petroleum mono-energy by electrical mono-energy."[6]

In pushing for renewables, antinuclear thinkers and activists argued that what was at stake was not a choice between different energy sources, but a choice between different energy models and, ultimately, different forms of social organization. The best-known and most influential formulation of this idea was Amory Lovins' distinction between *soft* and *hard energy paths*.[7] Largely associated with nuclear energy and fossil fuels, hard paths were described as based on the use of nonrenewable stocks and large-scale power plants with high-risk factors. Lovins believed that hard paths undermined democracy: they marginalized citizen participation and reinforced interventionist mechanisms of bureaucratic control and the power of oligopolistic corporations.

In contrast, soft paths were based on renewable energy flows. Built on the modular aggregation of small-scale, locally owned, and relatively low-tech energy production plants (such as wind farms and solar panels), soft paths were flexible and adjusted to a differentiated demand. This was a frontal attack on existing energy systems, based on the provision of increasing quantities of bulk power to satisfy an ever-rising demand. This hard-path logic had treated demand for, and the unfettered provision of, energy as a proxy for economic growth and the improvement of "quality of life."[8] Soft paths instead focused on reducing energy demand and matching end-use energy needs with energy quality, distribution, and scale—for example, by using thermal energy instead of electricity for heating or by producing an appropriate quantity of energy where it was to be consumed. If hard paths represented the reproduction of centralization and concentration, soft paths were to translate into the creation of *distributed energy systems*, a much-debated concept that ultimately indicates, in Alanne and Saari's words, "that energy conversion units are situated close to energy consumers, and large units are substituted by smaller ones."[9]

Lovins's well-known characterizations of hard and soft paths gave energy sources a pivotal role, implicitly assuming that the outcome, or path, would precipitate out of the use of specific energy sources. Where oil and gas can only be found in large quantities in certain parts of the world and nuclear plants require high technological sophistication and a large immobilization of capital, solar irradiation and wind are abundant, fairly evenly distributed, and can be harnessed with low-tech instruments. In other words, renewable energies would develop through and produce a soft path; their natural characteristics prefigured it. This ruptural assumption was to have a huge influence, as Sergio Oceransky explains: "For decades, the analyses and scenarios developed by most renewable energy advocates have assumed that the transition to renewable energy would result in decentralized and community-controlled renewable energy systems. The main reason behind this assumption is that renewable energy sources are decentralized by nature and easy to obtain by anyone with access to technology."[10]

Yet this proved incorrect, and the evidence against it piles up in parallel to the extension of renewables, showing that there is no necessary correspondence between renewables, decentralization, and democratization. In many parts of the world, wind energy production is increasingly concentrated in particular locations and centralized into fewer hands. Technological advances—especially the development of larger wind turbines and off-shore wind generation—have put the technology largely out of reach for ordinary citizens. And in diverse parts of the world wind energy has triggered, in the words of Oliver Hinkelbein, a series of "wind rushes" replicating the mining logic that Mumford described.[11] The real debate around energy transitions is not about shifts in technologies and resources, but about the structures that govern the energy system in specific historical contexts.

Green Wind, Black Wind

Wind energy development can follow a soft path or a hard one. It can be the spark of an alternative energy model or simply function as an alternative energy source to give continuity to the existing energy system. Frede Hvelplund, a long-term advocate of wind energy, has expressed this double possibility by distinguishing between *black wind power* and *green wind power*.[12] Green wind power is characterized by local and regional participation in wind power planning and ownership of wind turbines, reducing the distance between consumers and producers, and adjusting production to heterogeneous end uses. In contrast, black wind power is characterized by top-down planning combined with distant ownership in the hands of large companies and investment corporations. It is fully integrated within the long-distance electricity grid, producing bulk power for a homogeneous, growing demand. If green wind power promotes citizen participation, black wind power extends the centralization of the energy system and of decision making.

In a comparative study of five European countries, Joseph Szarka observes a strong correlation between black paths of wind energy development and policy frameworks that privilege economic over environmental goals.[13] Typically, these policy frameworks adopt some combination of two theoretical paradigms: natural capitalism and ecological modernization. The first is based on the idea that smart business strategies can solve environmental problems "at a profit," and it is perhaps unsurprising that Amory Lovins became one of its main advocates.[14] Ecological modernization is, in Maarten Hajer's words, "basically a modernist and technocratic approach to the environment that suggests that there is a techno-institutional fix to the present problems."[15]

Both paradigms are easily compatible, coupling support for business-as-usual models with technocratic management, altogether suggesting that market solutions and technology fixes can guarantee sustainable development, an oxymoronic term referring to ecologically sustainable economic *growth*.[16] In arguing that economic growth and environmental protection can be reconciled if subjected to expert administration, this perspective promotes what Slavoj Zizek calls a postpolitical understanding of politics—"as the art of expert administration, as politics without politics."[17] Indeed, what is at the root of the debate between soft and hard models, green and black wind power, is a clash between an understanding of energy as a technical problem to be addressed by experts within the framework of existing patterns of accumulation and class relations and, on the other hand, an understanding of energy as a political field of struggle from which a redefinition of the whole framework of production is a potential outcome.

For Hvelplund, the main problem with black wind power is that it foments popular resistance, whereas green paths are more easily accepted. All over the world we can find examples of popular movements opposing wind farm development. This grassroots opposition tends to puzzle policy makers and the wider public, who tend to dismiss it as the expression of shortsighted, NIMBY behaviors impeding the construction of a greener future.[18] Nonetheless, this explanation is too simple. Indeed, a growing number of ethnographic case studies make evident that resistance to wind farms closely correlates with the involvement of large transnational, foreign corporations, the existence of privatizing policies, the absence of local ownership, centralized planning, and lack of community involvement. Furthermore, several authors have pointed out that the fact that wind energy development tends to cluster in peripheral geographical areas, where higher prevalence of poverty and health risks combines with a lack of political capital, gives wind energy an extractive character, and foments the emergence of opposition movements.[19] In sum, grassroots resistance to wind energy development should not be viewed as a particularistic barrier to the birth of a post-carbon future, but rather as part of a broader, long-term struggle to define this future.[20]

A Missed Opportunity

In this section I will trace how wind development evolved in Spain from the early emancipatory vision to a form of green capitalism dominated by large corporations—that is, what wind energy could have been and what it has become. I will do so by describing the history of Tecnolia, a cooperative wind turbine manufacturer created by a group of radical engineers in the early 1980s that would eventually become part of a transnational energy corporation.[21] As we will see, the transformation of the company and of the direction of wind energy development runs in parallel with, and has in part provoked, a shift in the mainstream of the Catalan and Spanish environmentalist movement, which has left behind its radical origins and embraced more pragmatic positions.

"That Says It All"

Alstom Wind is a major wind turbine designer and manufacturer branch of Alstom, a French multinational with a strong presence in the global energy market, specializing in the manufacturing of turbines for hydro, coal, gas, and nuclear power plants. At Alstom Wind's international headquarters, in an old, reconverted industrial building of Barcelona, five hundred employees design turbines and wind farms. The company owns manufacturing and assembly plants in Spain and elsewhere; it develops wind farms, generally with the aim of selling them to electrical utilities; and it provides operating and maintenance services to wind farms that use their technology. It is headquartered in Barcelona because it was created after Alstom's acquisition of Tecnolia, a company founded by activist antinuclear engineers, for 350 million Euros. In 2015, two years after my interviews, Alstom was purchased by General Electric.

In 2013, I visited Alstom Wind's headquarters. Despite the nuisance of security checks, I quickly noticed a certain informality that contrasted with the corporate atmosphere I had encountered at electricity utilities throughout my research. After surrendering my national ID card in the reception area, I was met by Xènia, the dynamic engineer with whom I had arranged an interview. At the elevator bank, she said, "You may not have noticed it, but we are a bit agitated here these days. We had a strike a few weeks ago, and there is another one coming. Three factories have already been closed. Everyone's very upset."

A few months earlier, Alstom's management had proposed a labor-restructuring plan that would eliminate more than three hundred jobs. One union member had told me the factory closures and job redundancies were to be blamed "in part on the minister"—that is, they were partly a consequence of the elimination of subsidies—"and in part on the company, which is taking advantage of the situation to close off things here while they expand elsewhere."

Xènia said it more politely: "We keep selling turbines in other parts of the world, such as Brazil and the US, but these countries often demand that turbine manufacturing take place there, so the Spanish factories close. We are okay here [in the headquarters], because we are engineers and designers, but there is always the fear that they will relocate us, too. After all, this is a French multinational, you know what I mean."[22]

Later, after Xènia and I had chatted for a while, Marc, a younger engineer, joined us. Xènia had been hired while the company was Tecnolia, but Marc joined after the transition to Alstom Wind. We had a lively hour and a half long interview, returning, at the end, to "agitation" and the wind sector's problems.

Xènia initiated an animated dialogue: "If the wind companies only think about making money, that's not the right way. [Wind] engineers care about the industry; they love it. I went into renewables because I believed in them, because

I wanted to do something for the environment, and also because I am an engineer and I can make a living at it. I like being motivated in my job and I care about the environment."

Marc interceded, "This is also true in my case, but I am not sure this is representative of the people who have joined the sector in recent years. There's been a [wind] bubble, and that attracts a lot of people, all kinds of people."

Xènia replied, "Well, I still think there is this sentiment, especially in what was Tecnolia. The people working there were there because they wanted to do something for the environment, and the proof is that those who left have continued working on environmental issues. They have a kind of love for renewables. Now that we are Alstom maybe we are a bit disenchanted, but in the past it was very obvious, people came biking to work to avoid wasting gasoline. It was a cooperative, people felt it, and we had to do something for the environment."

Marc spoke again: "Well, that pretty much sums it up. We were a small Catalan co-op and now we are a French multinational. That says it all. Now we are part of a large company that makes all sorts of things, not only wind, it also has nuclear plants and many other things. And in a way we're lucky with that, because in this context, on our own, we might not be afloat."

Antinuclear Origins: Radical Technology

A few weeks before my interview at Alstom Wind, I had visited Xevi, one of the original cofounders of Tecnolia. Although he left the company in the 1990s, Xevi was still in touch with former coworkers and very involved in the wind sector, participating in some wind projects and working to spread the word about the virtues of this energy. Xevi, now a retired university professor, is well known in Catalonia, where he has been one of the most visible faces of the environmental movement and several green parties. As is the case with the vast majority of wind energy pioneers, his trajectory started in the antinuclear movement. He first became involved in the 1970s while studying nuclear engineering in Barcelona:

> We were trying to give a hand to the mayor of Ascó and all that, but then what happened is that at the same time a series of multinational companies, such as Chevron, obtained permits for uranium prospecting and mining in some parts of Catalonia, one of them the county I am originally from. That spurred some resistance and some of those who were in opposition came to see me. I knew nothing about uranium mining, but I collected information, got in touch with people in the Four Corners [region] in the US, where there was all the indigenous resistance and all that, and the more I knew the less I liked it. So I started going through the villages to give talks on this matter. And the well-intentioned question that many people posed to me was, "Okay, but then if we don't have uranium and nuclear plants, where will we get energy from?"

It was the million-dollar question. That pushed us to go further ... and explore these new [solar and wind] technologies.

In the late 1970s, Xevi started to work on a dissertation on wind energy and became involved, along with most of the others who would later found Tecnolia, in an intense activist campaign promoting renewable energy. Their ideas were influenced by the work of a wide range of radical environmental thinkers, and by the antinuclear struggle and the political context of the *Transición* (see chapter 2).[23] Many had been involved in clandestine political parties. They organized conferences, proposed alternative energy plans, got in touch with Danish pioneers, and created a series of organizations, such as a wind energy subcommittee within the Catalan college of engineers and the more radical organization Tecnologías Alternativas Radicales y Autogestionadas (TARA, Radical and Self-Managed Alternative Technologies).

TARA, created in 1977, edited a series of publications that show how these pioneers were imagining renewable energy at the time. The central idea was using renewable energy as an instrument to build a new society. For these activists, there was no point in wondering whether renewable energy could power the urban-industrial complex of "modern society"—the goal was to transcend that society and the growth imperative on which it was based in order to march toward decentralization and *autogestió*. The conclusions of the 1980 TARA-edited volume *Energías libres* (Free Energies) made this point clearly:

> Alternative technologies demonstrate the tangible possibility of recovering *autogestió* in all its range and nuances. This is why it is so important that hundreds of small groups and communities ... appropriate this technology against an omnipresent technological totalitarianism. For most of the defenders of solar strategy, the new energies are not mere technical substitutes for capital's energetic irrationality. For the first time, specific technological proposals carry with them the project of a different society.... Alternative projects implicitly carry within them the defense of decentralization in the face of the dictatorship of concentration. Concentrating energy is concentrating production, and thus a way of subjugating local autonomies to the centers of decision making. A new way of producing, consuming, and working can only be sustained by new energy sources and technologies.[24]

For these activists, the concentration of production in large-scale plants and the centralization of decision making were the two legs of a power system built on positing citizens as dependent clients. In other words, decentralization, deconcentration, and *autogestió* were the opponents of corporate ownership, technocracy, and large-scale production facilities. Reversing these dynamics in the energy system was necessarily a political attack on the foundations of capitalist accumulation as a historical system of exploitation and domination.

Nonetheless, the members of TARA expressed an acute awareness that renewable energies did not automatically translate into social change; effecting this change was a matter of social relations of production, not a technological one. As Xevi wrote in 1977, wind energy could develop along a hard path, neutralizing its transformative potential: "The development of wind energy must be geared toward the creation of a technology that is appropriated [*propia*], self-managed [*autogestionada*], decentralized, of small or medium scale, not oriented to productivity alone, and able to overcome the technical and social division of labor. Otherwise, no matter how inexhaustible, free, and clean they may be, new wind technologies will be controlled by multinational companies, which would mean that we have not advanced much."[25]

The Danish Model

It is tempting to judge the texts of the TARA pioneers as utopian and ahistorical. Nonetheless, it can hardly be overemphasized that they were deeply influenced by a specific, very real historical experience unfolding contemporaneously in Denmark. Xevi stressed this point time and again: "When we created the co-op, I was not interested in the university establishment, I was interested in the professional and social experiment, in demonstrating that we could develop technology as the Danes were doing. The Danish were the model in that moment. We travelled to Denmark a lot, we invited them to give talks, and we were fascinated with these small groups of engineers, blacksmiths, and local activists. We [thought that we] should be able to do this at home too!"

During the 1970s and 1980s, the Danes developed the basis of modern wind energy technology to such an extent that the term "Danish concept" today refers to the three-bladed, upwind design turbine that has become the icon of green power. Yet behind this technological feat lay a complex social process and a whole model of wind energy development—the "Danish model."[26] Pervaded by a strong communitarian spirit, experimentation with wind energy emerged more or less independently in several locations in Denmark's Jutland peninsula. It brought about a coalition of grassroots organizations, university researchers, and sympathetic parliamentarians in a historical context marked by an acrimonious nuclear debate and the severe effects of the oil crisis.

Denmark's process of "green innovation" was rooted in two historical legacies. One was the memory of the "first golden age" of Danish wind power, an expression that refers to the locally produced windmills—used for milling, land reclamation, and electricity generation—that proliferated on Danish farms during the early twentieth century, but ended in the 1920s as the extension of the electrical grid rendered small windmills obsolete. The legacy of the *folkehoegskole* (folk high school), a nineteenth-century movement around the idea of bringing enlightenment to rural citizens through practical education, also played a crucial role. Indeed, in the 1970s, one of the first large prototypes of a wind turbine was built and installed to serve the energy needs of a folk high school.

Autonomy, communitarianism, and local ownership shaped modern wind power in Denmark. Until the year 2000, Danish wind energy development was characterized by cooperatively owned, small, isolated wind turbines: "Twenty to forty households would typically own a mainly bank-financed wind turbine in their neighborhood. In 1990 there were around 100,000 relatively well organized cooperative wind turbine owners in Denmark."[27] This ownership model was linked to a pattern of spatial distribution. Local cooperatives owned single turbines—or, at most, a cluster of two or three turbines—largely serving local demand; turbines came to be scattered around the countryside with few in urban settings.

Early Danish wind development thus fits the concept of green wind power, where the Spanish case, as we will see, fits with the concept of black wind power: from the mid-1990s on, the latter was dominated by large investors, often electricity utilities, building rural wind farms of ten to fifty turbines to serve the needs of urban centers.[28]

From the First Turbines to the Wind Bubble

In 1980, spurred by the Danish experience, Xevi and seven other engineers and renewable energy activists formed Tecnolia (initially it was to be named Cooperative for Technological and Energetic Autonomy). The oil crisis of 1979 had severe effects on Spain, and the government, despite maintaining its support for nuclear energy, launched a series of initiatives to diversify energy production. These included subsidies for the restoration of small hydroelectric facilities and modest grants for experimental renewable energy projects. Tecnolia was formed ad hoc to compete for these grants, successfully applying for funds to build a wind turbine prototype. In 1984, this prototype was installed in the northeastern Catalan county of Empordà, part of the wealthy tourist area known as Costa Brava. It was the first wind turbine ever connected to the electricity grid in Spain. A month later, the first Spanish wind farm was inaugurated with five turbines, also an experimental project but in this instance promoted by the Catalan government and the public electricity company ENHER. This second project was installed in the same region, taking advantage of the strong north wind, called the *Tramuntana*, which traverses the area. A third pioneering wind project, a wind farm promoted by FECSA, came to the area in 1990. Since then, no other wind turbine has been installed there, and no trace of the three pioneering projects remains. According to Xevi, wind energy development did not take off in this period in Spain and Catalonia because political authorities were fundamentally pronuclear:

Those were difficult years. Between 1981 and the mid-1990s, surviving was a heroic act: we had a machine [the prototype] that worked, but the problem was selling it. We had some small projects in Albacete, Tarifa, and the Canary Islands, but we probably sold no more than thirty turbines. We combined turbine design and manufacturing with a bit of solar and a lot of awareness campaigns: we went to schools, did a video.... We did not distinguish between work and activism. We did activism and tried to live and make something useful that was coherent with our ideas.

This slow and sporadic development would experience a sudden shift in the mid- to late 1990s, when big electricity utilities and other large corporations, spurred by a series of legislative changes, especially feed-in tariffs (see below), started to invest in wind energy (see table 5.1). By the early 2000s, Spanish wind energy was experiencing an authentic boom, with dozens of new wind farms connected to the grid every year. Tecnolia's business expanded dramatically, though the cooperative lacked the financial muscle to compete in a hot market requiring growing volumes of investment. Tecnolia looked to Mondragón, the giant Basque cooperative, as an investment partner.[29] Xènia recalls: "[Tecnolia] started to sell turbines all over Spain. [There was] lots of work, I spent most of my time in the Spanish countryside. Except here in Catalonia, in fact, because not much was happening here." Catalonia was still an outlier. Although several factors—such as the regional government's lukewarm support to wind energy and the little interest that the utilities showed for the region—contributed to this situation, Xènia, Xevi, and all the developers with whom I talked blamed it on the "ferocious" and "intransigent" opposition that emerged in the region (see chapter 4).

In 2007, wind development was at its peak but the Spanish economy was showing signs of distress. Mondragón, whose business in manufacturing home appliances was faltering due to trouble in the real estate and construction sectors, withdrew from its alliance with Tecnolia.

Mondragón made a mistake. They admit it now. All the big players—General Electric, Siemens, and others—were entering the market, and it was more and more complicated to compete.... And the Catalan government was totally inept. They missed the opportunity to create a large wind industrial group. They should have made a strategic decision to do so, but you know the influence that oil and electricity companies have on the governments here. And Tecnolia was a small pain in the ass for them. [Tecnolia] is a legacy of the old ecologist movement, and when you demonstrate that you can be one of the best companies in the world, well, I guess it bothered them. Alstom bought Tecnolia, and the government did not move a finger. It was a perfect way to neutralize it. And Alstom found a golden opportunity: they were totally behind in the wind technology sector, and then suddenly they're in.

Table 5.1. Wind energy installed capacity in Spain, 1999–2014 (in MWh).

Year	Annual	Accumulated	Year	Annual	Accumulated
1999	720	1,433	2007	3,502	15,071
2000	906	2,339	2008	1,613	16,684
2001	1,156	3,495	2009	2,455	19,139
2002	1,505	5,000	2010	1,487	20,626
2003	1,160	6,160	2011	1,048	21,674
2004	2,280	8,440	2012	1,110	22,784
2005	1,551	9,991	2013	175	22,959
2006	1,578	11,569	2014	27	22,986

SOURCE: *Asociación Empresarial Eólica* (www.aeeolica.org).

The cooperative could not outlast the economic bubble. In 2008, it became part of a multinational corporation specializing in nonrenewable energy. Alstom's acquisition of Tecnolia, as we have seen, combined with increasing difficulties in the domestic market, led to the internationalization of the company's business activities. Indeed, wind energy development in Spain has remained almost completely frozen since 2013. Nonetheless, in August of 2015, Alstom Wind announced that, in partnership with Gas Natural, the largest gas corporation and third largest electricity producer in Spain, it planned to build three wind farms, totaling ninety MWh, in Terra Alta, already host to eight wind farms. These will be the first wind farms in Spain to operate without production subsidies of any kind. Later in 2015, General Electric acquired all of Alstom's energy transmission and generation assets for an amount close to 10 billion Euros. As a result of the purchase, Alstom Wind became GE Wind, in a further round of wind industry concentration.[30]

More Is Better

It is not difficult to see why for veteran Southern Catalan antinuclear activists who are critical of wind energy development, the thirty years of existence of Tecnolia/Alstom/GE evoke a feeling of missed opportunity. If we look back and compare Tecnolia's first wind project with Alstom Wind's latest proposal in Catalonia, the contrast is striking. In 1984, Tecnolia was a local cooperative trying to transform the world by installing a fifteen kWh (0.015 MWh) turbine near a major international tourist destination; in 2015, two large, multinational energy corporations with core businesses in nuclear energy and fossil fuels plan to build a ninety-MWh wind farm in Terra Alta, the poorest, most peripheral county in Southern Catalonia.

Nevertheless, Xevi disagrees that this type of wind energy development represents a missed opportunity. Indeed, he dismisses the idea that wind energy is currently concentrated and centralized: "Who says that a 50 MWh wind farm is a big power plant? I don't see it as centralized. A nuclear plant is large-scale, this isn't."

Xevi is mildly critical of some aspects of wind energy development, such as the absence of turbines in cities and coastal areas and the dominance of large energy companies over the sector. He also thinks that wind developers have been shortsighted and greedy in their relationship with local inhabitants, and he complains that they are unwilling to establish mechanisms favoring some share of local ownership: "It was a strategic mistake of the companies. I told them time and again: keep your majority ownership but open five to ten percent to local investors. But no way, they wanted it all; this is their mentality."

Despite these criticisms, Xevi defends what he would call a pragmatic vision, advocating accommodation to the realpolitik of wind development, seen as good in itself, regardless of the path it takes. For him, more wind energy is always better, no matter how it is created; the real drama, the true missed opportunity, is the paralysis of the sector. Xevi sees the Southern Catalan opposition movements as complicit with this paralysis, and he has been at the center of public, acrimonious confrontations in Southern Catalonia. Old comrades of the antinuclear struggle have become irreconcilable foes in one of the more lamentable legacies of the wind bubble.

Xevi's attitude represents a large segment of the Spanish environmental movement that is not only numerically dominant but also hegemonic, largely monopolizing the environmentalist discourse. Indeed, the transformation of wind energy development has gone in parallel with the evolution of the mainstream environmental movement and its growing acceptance of market solutions within an ecological modernization paradigm. As activists like Xevi usually say, "Every megawatt of wind energy is a megawatt of oil that does not get burned"— even if actual research contradicts this statement.[31]

In 2014, the late Ladislao Martínez, a well-known veteran of the ecologist and antinuclear movements, published an insightful article on the contradictions at play in asserting that more renewable energy production is always better. The article starts by reviewing his contributions in some of the monographic issues edited by TARA in the 1970s, expressing surprise at the group's naiveté: "How different from what we imagined has it all been!" Martínez then offers his sober appraisal of what happened:

> Renewable energy has not been developed to satisfy the energy demands of the population, but rather to maximize the profits of corporations. It was not created in order to build a harmonic relationship with nature, but just to

simulate activity in the face of the environmental challenge of climate change. All this in a context where all other corrective measures (conservation, efficiency, greenhouse gas emissions reduction) were totally absent.... Yet, we should not forget the concrete history of wind power in our country. While companies that were not part of the electrical oligopoly promoted the first wind farms, wind energy only took off once the large companies of the sector jumped in. As a result, there is no public participation in the sector, and no local ownership of wind farms, and these circumstances nurture the idea that wind energy is "just another business."[32]

Martínez stresses that large corporations, with the traditional oligopolistic electricity utilities at the fore, have dominated wind energy development in Spain, imposing their economic rationality onto it. This hard path can be encapsulated in three characteristics: centralization of ownership and decision making in a handful of major corporations, a productivist understanding of energy, and concentration of production in a series of peripheral rural regions that functioned as areas of extraction.

Martínez laments the deactivation of wind energy's transformative potential. Yet, it is also clear in his piece that he sees corporate control as a necessary evil for bringing wind development and its environmental benefits to fruition: "The fact that wind power has become the main source of electricity in Spain in 2013 ... has only been possible because of the decision to proceed to a centralized implementation concentrated in some of the best sites."[33] Oligopolistic capitalism, he suggests, may not be the best option, but it was the only available one, and it worked. The takeoff was nothing short of phenomenal: from less than 1 GWh of installed capacity in 1998 to 22.7 GWh in 2012 (see table 5.1). In 2014, wind accounted for 20 percent of the electricity produced in Spain, narrowly second to nuclear at 21 percent. Globally, only Germany and Denmark can offer comparable figures.[34] It is tempting to conclude that Spanish wind energy development is an example of the efficiency of market solutions for green problems. Wind energy development created a new business sector that facilitated the internationalization of Spanish utilities and manufacturers. At its peak, this sector accounted for 1 percent of the gross domestic product and sustained a large number of jobs in a country that suffers from endemic unemployment.[35]

For well-intentioned activists such as Ladislao Martínez, the pros of the hard path largely outweigh its cons. True, the Spanish electric system has not been democratized, but pro-market measures and the involvement of the electric utilities have enabled the growth of a sector with unmatched environmental benefits and the promise to transform and strengthen Spain's economic base. This argument, however, has a critical flaw: the benefits of wind energy development have been rather modest.

Table 5.2. GHG emissions, 1995–2014 (in millions of tons, CO_2 equivalent).

Year	EU	Spain	Year	EU	Spain
1995	5,399	332	2008	5,179	422
2000	5,283	395	2010	4,912	373
2005	5,348	450	2012	4,692	369
2007	5,342	453	2014	4,419	342

SOURCE: *European Environment Agency* (http://www.eea.europa.eu/data-and-maps/data /data-viewers/greenhouse-gases-viewer).

Consider the environmental benefits: during the wind bubble, when Spain was increasing the energy produced from the wind in absolute and relative terms, the country's environmental indicators worsened. The dramatic rise in greenhouse gas (GHG) emissions contrasted with the broader pattern in the EU, jeopardizing Spain's capacity to meet the targets of the Kyoto protocol, which established a maximum increase of 15 percent between 1990 and 2020 (see table 5.2).[36] Revealingly, GHG emissions in Spain only improved after 2008, as a result of the decrease in economic activity (particularly housing and construction) that signaled the beginning of the economic crisis.

Wind energy's economic benefits followed the opposite path. During the miracle, these benefits were very tangible, especially in terms of jobs, but with the crisis, when these jobs were most needed, they disappeared.[37] Ulrich Jochimsen, a long-time advocate of decentralized renewable energy has warned: "Those wishing to see and contribute to a just and sustainable ecological and social development ... would be advised to sever ties with the unscrupulous profiteering which is being carried out behind the camouflage of deceitful labels such as 'sustainable' and 'renewable.'"[38]

The worsening of environmental indicators and the fact that the sector could not detach its fate from an economic crisis triggered by the housing bubble suggest that, despite its disruptive potential, wind energy development did not represent an alternative to the dominant logics of accumulation, but rather a continuation of them. Wind energy development not only integrated within the existing energy system, thus being subjected to the interests of the electrical oligopoly and to the economic rationality that has historically governed electricity systems. It also was the creature of the Second Spanish Miracle, bearing all its distinctive features, ultimately one more bubble in a bubble-led economy.

Wind and the Second Spanish Miracle

Between 1995 and 2007, the Spanish economy grew at an average rate of 4 percent per year, a rate bettered only by Ireland within the Eurozone, while creating more than half of all new jobs in Europe.[39] The Second Miracle was, however, much more than an economic bonanza; it was an overarching process with profound influence over the entire social fabric and territory of the country. It involved the consolidation and partial reconfiguration of the patterns of accumulation, class relations, institutional architecture, environmental dynamics, and power structures of the country. The extent and depth of the crisis that followed, both at the economic and political levels—with unemployment figures far above 20 percent and unparalleled levels of mistrust toward the country's institutions—perfectly mirror the capillary extension of the miracle.

The Second Miracle reflected key continuities with the First Miracle of the 1960s and 1970s (see chapters 2 and 4). Although it facilitated the entry of new actors—starting with the political cadres of the democratic parties—it also perpetuated a form of capitalism whose intimate texture is starkly described by journalist Maruja Torres: "Budget overruns as a form of public administration, [illicit] commissions as a social aspiration, immediate profit as the promise of paradise."[40] As I discussed in chapter 4, corrupt deals, the reinforcement of a rent-seeking oligarchy, and the overgrowth of the real estate sector left a profound imprint on the Spanish territory, whose degradation came to embody the excesses of the Second Miracle and to spark largely uncoordinated, widespread resistance to these "territorial assaults."

In the remainder of this section, I will describe how wind energy development was inserted into and became part of the Second Miracle. While a detailed analysis of this period is beyond the scope of this book, here, I focus on what I understand to be its three main, interrelated characteristics as evident in the wind bubble. First, over the course of the miracle, the oligarchic structure of power centered on a corporate core was reconstituted and consolidated. Second, widespread cronyism and corruption filled in the interstices between political and economic power in order to capture politically created surplus opportunities and consolidate class divisions. Third, the country experienced a transformation in the structure of accumulation that increasingly tied profits to forms of monopoly rent extraction, led by construction growth and real estate inflation. Analyzing the integration of wind energy within the Second Miracle therefore requires a displacement of our position within the wind electricity commodity chain, from turbine manufacturing to wind farm development, which was the ultimate driver of the expansion of the wind energy sector. After all, without new wind farms, no turbines would be sold.

Oligarchic Reconstitution of Power

Around two thirds of Spanish wind energy production is owned by large electric utilities (whose monopoly in nonrenewables is almost total) and, to a much lesser extent, large construction companies. Since the beginning of the crisis, this percentage has shown a modest yet unmistakable tendency to increase.[41] The rest is in the hands of mid-sized independent developers (which typically own one to ten wind farms), investment funds (often linked to real estate) and turbine manufacturers. The share of electric utility ownership is far larger than in other European countries, such as Denmark and Germany, with a similar rate of wind energy development. A comparison with the German case is especially revealing, for the regulatory framework adopted in Spain was inspired by German legislation and experience.

The takeoff of wind energy can be dated to 1998, with passage of a decree to foment the development of renewable energy. The legislation was inspired by the German feed-in law, approved in 1991, which established a subsidy for production or feed-in tariff (FIT). Beyond certain national specificities, two elements are central to FIT systems: they establish a preset price or tariff for renewable energy and, importantly, require the grid operator to give renewable energy priority access to the grid (i.e., in case of excess supply, renewables displace other energy sources).[42] While EU institutions and certain state members have tended to favor competitive market mechanisms to promote renewable energies, FIT systems have been more successful at expanding renewable energy production: by guaranteeing the sale of electricity at a set price, they make returns more predictable and thus incentivize investment.[43] To be sure, wind energy involves a significant investment: the ballpark figure in the sector is 1 million Euros per installed MWh. This regulatory system also made wind energy investment very profitable: typically, a wind farm with a life of thirty years covers its costs within the first fifteen years of operation.[44]

Yet, there was a crucial difference between the Spanish and German regulatory frameworks. The German feed-in law imposed regulatory barriers to facilitate the entry of new actors into the electricity system. It prohibited large, existing utilities from owning more than 25 percent of any renewable energy project and from investing in the regions where they had monopolistic control (thus limiting their political influence). In contrast, the Spanish legislation adopted what some supportive commentators euphemistically called a "collaborative approach" between government and large utilities.[45]

While German utilities unsuccessfully challenged the country's feed-in law in court, Spanish utilities engaged in wind energy development and established a leadership position within the sector. It is clear that the Spanish utilities recognized how the upholding of the German law sent, as one developer put it to me, a clear message: "we cannot stop it, so let's join in; better us than someone else." Thus, for instance, whereas in 2005, Iberdrola owned 3,500 MWh in Spain, in Germany the utility with the most installed wind capacity, E.ON, owned a mere 224 MWh.[46] Additionally, in contrast to the German context of stable electricity demand, meaning that new wind energy sources would displace nonrenewable electrical generation, Spain's growing electricity demand meant that wind energy generation could expand without jeopardizing the utilities' nonrenewable interests.

To understand more about the Spanish electricity corporations and their reaction to wind energy, it is necessary to look at the broader regulatory changes that shook the sector during the 1980s and 1990s. At the end of chapter 2, I described how the nuclear program had left the Spanish electricity sector in disarray. In 1988, the government, then in the hands of the social democratic party PSOE, introduced a new regulatory framework called Legal Stable Framework (Marco Legal Estable or MLE).[47] The MLE strengthened the state's control over the electrical system, regarded as a basic good with strategic importance for national development. Thus, through the MLE, the state nationalized the electrical grid, determined the price to be paid by consumers, and guaranteed moderate, predictable profits for the electricity corporations. These limits on profitability, together with the dire financial straits of the electricity corporations, help explain why investment in new power plants was minimal in this period.

Yet, the MLE was short-lived. In fact, from a historical perspective, the intervention of the state may be interpreted as an operation oriented to allow the sector to repair its damaged financial position in order to undertake a new round of expansion. Throughout the 1990s, to an extent in response to EU guidelines, the state began to introduce a series of liberalizing measures. In 1997, it approved a new law for the liberalization of the electric sector (Ley de liberalización del sector eléctrico). This high-profile legislation was a star project of the Partido Popular (PP), the conservative party that had reached the government embracing the neoliberal principles of free competition and unfettered markets.

The law was meant to showcase the benefits of substituting market mechanisms for state control: the government maintained ownership of the long-distance grid while creating a wholesale market for electricity retailing (the sale of electricity to end consumers) and deregulating generation (the production of electricity, historically the key and most profitable phase of the electricity commodity chain). In this way, the law theoretically opened the sector to new actors, but it also surrendered the state's capacity to orchestrate the development of electricity production capacity, an outcome that would have far-reaching consequences. According to the government, the law would also foster cheaper electricity, as Josep Piqué, the Minister of Industry, explained in an open column in the (then) left-leaning newspaper *El País*: "The process of liberalization [will] allow the reduction of electricity prices [for consumers] and energy costs for corporations, thus bolstering their competitiveness and making our country one of the most open in Europe."[48]

The fact is that the liberalization of the electricity sector, far from fomenting competition among a larger number of agents, strengthened the oligopolistic core of the sector. This seemingly contradictory outcome stemmed from the government's facilitation of large corporations that could compete in an increasingly globalized market.[49] Thus, throughout the 1990s, the government oversaw a series of mergers and acquisitions that further concentrated the sector: the members of the oligopoly were reduced from about a dozen in the early 1980s to the current big four—ENDESA, Iberdrola, Gas Natural, and EdP, known in Spain as *the oligopoly*. In addition, the Spanish government privatized the only state-owned utility, ENDESA, as part of a larger process of privatization of public corporations undertaken by the PP government.

These corporations, together with the electric oligopoly, banks, construction companies, and, to a lesser extent, major tourist operators, would form the increasingly financialized, renovated core of the Spanish economy, the new "brotherhood of cement" (see chapter 2) pushing for growth in Spain and beyond. Indeed, in the late nineties these core corporations started to internationalize their business activities, initially taking advantage of the cycle of privatization of public services and companies in Latin America.[50] One last characteristic of this core of Spanish capitalism is the high degree of interpenetration among corporations, mostly through financial participation and board positions, and with the political sphere, the infamous "revolving doors," especially well oiled in the electricity sector.[51]

The involvement of electricity utilities and large construction companies in wind energy development makes intuitive sense in the context of the oligarchic reconstitution of economic and political power, in which the "collaborative approach" is a reflection of the broader intertwining of political and economic power. For the utilities, renewable energy represented an opportunity for domestic expansion with little danger to their existing, nonrenewable production. The success of wind energy in Spain also offered utilities a ready opportunity for replication and adaptation in select foreign markets. Finally, wind energy development provided the utilities an opportunity to improve their reputation, bolstering a new corporate image as "renewable leaders," even when their core business remained solidly anchored in the nonrenewable sector.[52]

Despite the dominant position of the utilities, one quarter of Spanish wind energy was initially developed (and owned) by independent, middle-sized developers. While we can find these independent developers in all Spanish regions, or *comunidades autónomas*, their presence was especially notable in those *comunidades autónomas*, such as Catalonia, where the oligopolistic utilities were less active in pursuing wind development. Given its heterogeneity and the lack of academic research on these developers, it is not easy to offer a detailed picture of this class fraction, but my fieldwork suggests a series of common characteristics. The mid-sized independents tended to be engineers, and often possessed family links to the traditional electricity sector. Many owned small construction or engineering companies specializing in public works, and their involvement in wind energy emerged as an extension of their business.

At the subjective level, these developers see themselves as entrepreneurs and honest businessmen, a position they define in opposition to three other figures: the "activist," the "monopolist," and the "opportunist." Some developers possess a long trajectory of involvement with renewables, but they have no past involvement in the antinuclear or environmental movements and do not conceive of their wind energy work as activism. They manifest a visceral disdain for the monopolistic practices of the utilities, which they contrast with their own entrepreneurship. Finally, the developers insist on distinguishing themselves from opportunists, their term for certain politically well-connected persons who acquired wind farm authorizations (and the guarantee of feed-in schemes) solely to sell those rights to large utilities.

In practice, however, the position of these independent developers is not as clear-cut as they would believe, and it is not always easy to distinguish their position from the "monopolists" and the "opportunists." The story of Gerard, one of the first wind developers to arrive in Southern Catalonia, illustrates the fuzziness of those distinctions. In the early 1980s, Gerard founded an environmental engineering company specializing in water treatment and the restoration of small hydroelectric facilities. Over time, he built extensive connections with other European entrepreneurs in the environmental business sector. In 1994, when wind energy was in its infancy in Spain, he received a phone call from a European friend who had recently moved to the United States to work for Zart, a US renewable energy company (see chapter 4). This friend told Gerard that Zart's owner was interested in knowing more about the wind sector in Spain, giving rise to what Gerard calls a "symbiotic relationship" that extended for about a year over the phone: "I taught him about the regulatory framework here and he taught me about the wind business."

Having discovered an interesting opportunity for the expansion of his business, Gerard registered half a dozen of wind farm projects in Southern Catalonia. "Then one day the American called me again: he wanted to buy all our [projected] wind farms. We hadn't built anything, but we didn't want to sell. I later found out that at the same time he was buying things [wind farm projects] in Greece and the UK and other parts of Europe. He made a package, put a ribbon on it, and sold it all to Enron [in 1997]." In 1999, Gerard finished construction of a wind farm in Southern Catalonia, the first one in the region and the only one he was ever able to bring to fruition.

Capitalismo de Amiguetes

Since the start of the crisis, and thanks to the invaluable work of journalists, prosecutors, social movements, and left political parties, the Spanish public has seen an unending parade of corruption cases involving politicians and private businesses. Spaniards have coined the expression "buddy capitalism" (*capitalismo de amiguetes*) to describe these scandals, the merit of which consists of presenting these shady deals not as malfunctions of the political-economic system, but as part of the very fabric of it. Behind the corruption scandals we find a series of clientelist structures and relationships blurring the line between public and private. Most involve the banking and construction sectors, but renewable energy has not been immune to corruption, and these shady deals tell us much about the accumulation structure and class implications of the kind of wind energy sector that emerged during the Second Miracle.

In my fieldwork, I encountered widespread suspicions of corruption and cronyism among ordinary Southern Catalans: "Why does the cousin of the mayor have a turbine on his land?" "Isn't it a nice coincidence that the minister has approved a wind farm owned by the corporation he used to work for?" "How come all the developers that have concessions around here are close to the ruling party?" While most of these accusations remain undocumented if not unfounded, my interviews with developers and media coverage of certain recent corruption scandals make evident that beneath the generalized suspicion there is an objective basis.

Indeed, if ordinary citizens tended to phrase their suspicions in vague terms, the complaints of medium-sized wind developers were more specific. Alfredo, a veteran developer, received me in a luxurious office in Barcelona with the words, "I was sort of a hippie, this is why I started with this. I was doing an MA in California in the 1980s and saw the wind experiment that was taking place there and thought it was great." No longer a hippie, Alfredo explains why, despite being Catalan, he has barely done any business in the region: "I have only built one wind farm in [Southern] Catalonia. And it took me eight years to get the permit. You know why? Because [the government] asked me for an [illegal] commission, 3 percent. I didn't pay it, and I've built most of my wind farms in Galicia, no one asked me for money there. Then the government changed [in Catalonia], and the wind farm got the authorization in a few months. We built it quickly, it is up and running and very profitable."

In 2014, the media aired a corruption scandal affecting the concession of wind farm authorizations in the *comunidad autónoma* of Castilla y León—the Spanish leader in wind energy production—which shows the deep interconnections between monopolistic and opportunistic practices. The Spanish revenue agency uncovered a long-standing clientelist network of politicians, bureaucrats, and "local businessmen" who had presumably received payments exceeding 110 million Euros from electricity companies in exchange for easing and speeding the authorization of wind farms.[53] The network initiated the bureaucratic processing of wind farms, but had no intention to build them. Instead, they sold their authorizations to real wind developers, most often large utilities. The main suspect in the investigation now lives in Poland, where he owns—surprise, surprise—a real estate company. The regional government has admitted that it did favor local businessmen, but says that it did so for the sake of "regional interests." The class dimension of these shady deals, dressed as "regional interests," is obvious. Although extremely variable, land leases—the amount wind developers pay the multiple local owners of the properties—typically represent, combined, around two hundred thousand Euros/year for an entire fifty-MWh wind farm, a laughable sum when compared to the profits made by corrupt regional elites.

Manel's business projects were directly affected by the workings of the corrupt network. This Catalan engineer and developer had licensed, in the early 2000s, several wind farm projects in Castilla y León. "But then—he explains—there were elections and in came a new regional president [of the *comunidad autónoma*]. It was the same [political] party, but he wanted to put in his friends. And that was it; we had to leave. The good thing is that thanks to the fact that we had partnered with a local bank [*caja de ahorros*] with good [political] connections we could get a good price [for the registered projects]."

Manel's wind businesses in Catalonia have not been very successful either. Although, at the turn of the century, he was able to register a dozen wind farm projects, he was only able to build and operate two. In this case, however, he does not blame political interference, but opposition groups and platforms, to which he expresses unconcealed resentment: "Why couldn't I do more wind farms? Well, the administration was not very helpful, they didn't believe in it, to be honest. But mostly because of the ferocious, visceral opposition that I encountered from the preservationist world—and I say preservationist because I don't think they are environmentalists.... They just make excuses, they talk about the eagles, and the birds, and all sorts of excuses. This is what they are, excuses."

A few months after we talked, media outlets revealed an important omission in Manel's story. Reportedly, in 2002, the vice president of the Catalan government sent a letter to the Catalan Minister of Environment saying that he thought that four wind farms that had been paralyzed for environmental reasons should be authorized.[54] Nine days later, these farms, two of which belonged to Manel, received a green light. The news suggested that this decision was linked to the fact that Manel was a co-founder of Convergència i Unió (CiU), the political party that controlled the Catalan government from 1980 to 2003.

In openly denouncing corrupt practices, independent developers such as Manel and Alfredo wanted to distance their businesses from them, presenting themselves as entrepreneurs who—in contrast to monopolists and opportunists—did not rely on political favors. Yet, the fact of the matter is that, given the nature of wind energy development during the wind bubble, the activity of these developers was heavily dependent on good relationships with the regional administration. Moreover, they relied on regional banks (*cajas de ahorro*), in which regional administrations were heavily invested, to finance their projects.

Willingly or unwillingly, medium-sized wind developers tended to find themselves replicating, at the regional level, the oligarchic structure of power of the utilities at the national level. Ultimately their activity was inserted in and dependent on a complex network of political obligations that constituted the fabric of the Second Miracle as a class-inflected structure of accumulation. The wind sector was, in fact, inseparable from these clientelist structures for two interrelated reasons. First, obtaining governmental authorization and bypassing local opposition was the crucial chokepoint in the wind business (as I will expand on in chapter 6). Second, wind development largely became a branch of a construction dynamic that provided the fertile ground for the generalization and hypertrophy of these clientelistic mechanisms.

Construction and Growth

The economic engine of the Second Miracle was the construction sector. Indeed, the Spanish miracle was built on an unparalleled expansion of what David Harvey calls the secondary circuit of accumulation, underpinned by abundant cheap credit.[55] By diverting investment to the built environment, the Second Miracle coexisted with, and in fact helped produce, the unraveling of the country's industrial fabric (i.e., the weakly competitive manufacturing base largely created during the *desarrollismo* years). Profits and growth were progressively de-linked from productivity, becoming, to a large extent, the product of different forms of rent appropriation, be it through massive private indebtedness, land valorization, or state-orchestrated transfers of wealth.[56]

Both the public and private sectors participated in this construction fever. The most visible manifestation of state involvement, often supported with EU funds, was the frantic activity of infrastructure construction: by the end of the growth cycle, Spain had the most kilometers of high-speed rail and highway per inhabitant in the EU.[57] Private construction activity had its paradigmatic expression in housing construction: between 2000 and 2005, Spain built as many houses alone as France, Italy, and Germany (with a total population five times that of Spain) combined.[58] The construction fever particularly impacted municipalities, the weak link in the administrative structure of the state. Historically underfunded but with large planning capacities, municipalities, especially those situated at the new frontier of ground rent valorization, became addicted to real estate growth. At the height of the miracle, more than half the budget of Spanish municipalities depended on real estate and construction activities.[59] Municipalities felt compelled to keep rezoning land to allow for new real estate development, even as their debt spiraled. This mutual dependence generated the perfect context for the corrupt maneuvering across public and private interests that we saw earlier.

The construction fever had two direct effects on the electricity sector: it drew new capitals to it and it increased electricity demand. The renewables sector became a coveted area of investment for construction companies, especially after 2004, when real estate and construction started to show some signs of exhaustion. This shift toward renewable energy occurred at all levels. Large construction companies became major stockholders of electricity utilities, some developing their own subsidiaries for wind energy development. Medium-sized construction companies expanded into wind farm development. Large individual fortunes created in the real estate business also sought refuge in the renewables sector, acquiring wind farms that were originally owned by independent developers. As Prieto and Hall argue, subsidies to renewables became a "massive government-endorsed transfer of funds from the public to investors."[60]

As this transfer between construction and electricity companies suggests, we cannot truly speak of the construction bubble affecting the electric sector; instead the electricity sector was part and parcel of the construction boom and its short-term, spiraling logic. This becomes even clearer when we look at the second effect of the construction fever on Spanish electricity. Indeed, electricity companies in both the renewable and nonrenewable subsectors benefited from the growth rates of the Spanish economy. Electricity demand increased by more than 30 percent between 2001 and 2008, outpacing gross domestic product (GDP), a result of construction's intensive use of materials and energy (see table 5.3).

Electricity companies responded by augmenting their productive capacity—building new power generation units. But revealingly, electrical productive capacity soon increased at an even sharper pace than electricity demand (66% vs. 32%, between 2001 and 2008).[61] Construction fed construction. Although the development of wind energy during the first decade of this century was formidable, it only represented part of the story. Indeed, the main drivers of this growth were state-of-the-art (combined-cycle) natural gas power plants, entirely owned by the utilities, which more than tripled their installed capacity between 2001 and 2008 (see table 5.3). Wind expanded in parallel to fossil fuels, making thus clear, as Prieto and Hall conclude, "that renewables [did] not displace fossil fuels but just added to the mix."[62]

These operations appeared to involve minimal risk. While the return on the investment in wind farms was guaranteed by the FIT system, the Spanish legislation stipulated that natural gas power plants would receive "capacity payments" when not in operation. The rationale for these payments was that the electricity system needed excess or backup productive capacity in order to respond to the increased potential for intermittencies that came along with the extension

Table 5.3. Electricity production in Spain by primary source, 2001–2014 (in %).[1]

Year	Hydro	Coal	Fuel, Oil	Nuclear	Cogeneration	Natural Gas	Renewables[2]	Total (in GWh)
2001	18.4	30.4	9.7	26.8	4.6	5.6	4.5	238,010
2004	12.3	28.8	8.0	22.6	8.8	11.6	8.0	261,307
2007	9.7	24.1	6.2	17.6	9.2	22.5	10.6	312,138
2008	8.3	15.8	6.0	18.6	9.2	29.5	12.4	316,850
2010	15.2	8.8	5.3	20.6	9.8	22.1	18.4	300,996
2012	8.1	18.5	5.2	20.7	11.0	13.6	22.1	297,876
2014	15.5	15.7	5.0	20.6	9.0	8.0	25.4	277,876

SOURCE: *La Energía en España* (Ministry of Industry, www.minetad.gob.es).

1. Totals may not adjust to 100 percent due to round up. For 2012 and 2014, the Ministry of Industry changed the form to aggregate data; the missing 0.8 percent for both years corresponds to waste treatment plants.

2. Includes wind, solar, biomass, and biogas. The share of wind is above 70 percent every year.

of wind energy, a stochastic, not entirely predictable energy source. However, the dramatic infra-utilization of combined-cycle power plants in the period after 2008 constitutes powerful evidence that their growth was excessive—fueled not by electricity demand but by the profitability of building power plants.

The expansion of energy production ultimately was a concrete manifestation of a broader construction craze, as Silvia, a young wind promoter working for a large utility, explains frankly: "It was the same with all the useless airports and highways that were built all over the country. The energy planning of the government was: I want renewables, and I'll give you these incentives until we get to the target of 22,155 MWh. And then everyone went crazy building wind farms! This is great, but if then you also give incentives for building combined cycles, and you don't look at what's going on, and keep giving permits… And now we have this whole mess." Let's then take a look to the "whole mess" that Silvia refers to, that is to say, to the circumstances that brought about the paralysis of wind energy development in Spain.

The End of the Bubble: Crisis and Debt

In 2008, the Second Miracle reached an abrupt end: credit evaporated, growth stopped, unemployment skyrocketed, and migration flows reversed. If triumphalism was a hallmark of the miracle, the crisis unleashed a profound, ongoing crisis of legitimacy in the economic and institutional structure that had emerged out of the *Transición*. Overcoming the crisis will probably have to wait for a new cycle of credit- and construction-fueled growth, whereas the outcome of the political crisis is being determined through a combination of electoral contests and street struggles (marked by the afterlives of the 2011 *indignados* revolt and the surge of Catalan self-determination demands).[63]

The destruction of the natural landscape—largely to make place for the second residences of Spanish urbanites and northern European tourists—was the main symbol of the ruthlessness of the miracle during its peak years; in its place, we now see a parade of corruption scandals that capture the indignation of large sections of the Spanish population. "This is not a crisis; it is a fraud," chanted the *indignados*. Scandals have converted the heroes of the miracle into the schemers of the crisis—witness the 2015 money laundering arrest of Rodrigo Rato, the former Minister of Finance and managing director of the International Monetary Fund often called "the architect of the [Second] Miracle." In parallel, the view that the miracle was a mirage, perhaps a nightmare, spreads. Whatever it was, it was class-inflected, as the cascade of housing foreclosures and the increase in poverty rates so poignantly indicate.

Electricity has become a tangible presence in the political debates on and around the crisis. For most citizens, electricity companies represent all that was wrong with the Second Miracle: fabulous profits, shady deals, oligopolistic practices, and the revolving doors between political offices and corporate corridors. At the same time, the combined effect of skyrocketing electricity prices—now among the highest in the EU—and decreasing household income has led to alarming rates of "energy poverty." This expression, barely known before the crisis, has become a commonly used term and a commonly experienced reality for many Spaniards. According to a 2016 study, 11 percent of Spanish households cannot adequately warm their houses during the winter, and 3 million people delay the payment of electricity bills (a 73% increase since 2008).[64]

The degraded reputation of electric utilities is no less tied to their role in paralyzing renewable energy development. The battle to reform energy policy in Spain must be viewed in the context of the broader European field. Indeed, since 2011 there has been a concerted effort by Eurozone's main electrical corporations to eliminate subsidies for renewable energy, presented as the driver behind soaring electricity prices.[65] This is nothing less than a demand for the EU to revise its approach to climate change and renounce its objective of meeting the targets set for 2020 by the Kyoto agreements. The corporations argued that respect for the environment should not come at the expense of competitiveness and the security of supply, especially in a time of economic crisis. This is the same line of argument that big electrical corporations deployed in Spain. Yet in this country, the corporations had a particular element in their favor: the "electrical debt."[66]

This debt has its origin in the concept of "electrical deficit" that emerged in 1997 with the so-called liberalization of the electric sector. As we saw earlier, this liberalization was a banner project of the PP government meant to prove that market fundamentalist recipes based on liberalization and privatization benefited citizens. The need to make the reform appear successful led to the invention of the deficit: government and electrical corporations agreed to mark down the consumer price of electricity, thus allowing the government to claim that liberalization improved efficiency and saved citizens money. The electrical companies financed the ensuing deficit, and the government promised to repay them in eight years, thus displacing the financial costs of liberalization to future consumers. In exchange, the government stipulated that the oligopoly would receive compensation for the monetary costs that the transition to a liberalized system would supposedly cause them. Thus, in the ten years following the liberalization of the sector, Spanish electricity consumers paid the utilities an extraordinary sum (upward of 10 billion Euros) via a surcharge in their electric bills.[67]

Altogether, the electricity policy framework created with liberalization ran quite smoothly up until 2006. Electricity prices were kept low, deficits were small, and the development of renewable (and nonrenewable) energy was impressive. But in 2008–2009, a perfect storm gathered: electricity consumption fell, subsidies grew, and deficits started to get out of control. Making things worse, the government took charge of the deficit and securitized part of it, converting it into a debt with interest that started to push electricity prices up. In 2013, this debt reached an astonishing 26 billion Euros. Because it is supposed to be repaid through future electricity bills, Spanish citizens and the whole electrical system are indebted to the electricity utilities for the foreseeable future.

The oligopoly mobilized this debt to lobby the government to initiate a regulatory reform of the electricity sector favorable to its interests. The end results of this campaign were a series of decrees, passed between 2012 and 2014 that eliminated the FIT system, bringing renewable energy development to a halt and sending the wind energy sector into disarray. The rationale that the government and the utilities offered for these reforms was straightforward: the electric system was on the brink of collapse due to a crippling debt caused by misplaced subsidies to so-called "immature technologies." It was thus imperative to eliminate those subsidies in order to restore the system's "financial sustainability"—even if this exposed the Spanish government to a long and growing list of corporate international lawsuits.[68]

We cannot know with precision how much truth there is in the argument of the utilities, for big electricity corporations and the major Spanish political parties have blocked all sorts of initiatives to audit the costs of the electrical system. Nonetheless, there are powerful reasons to doubt its veracity. Indeed, the main factor behind the escalating debt seems to be the defective design of the regulatory framework, which consists of a combination of competitive market mechanisms—what is called the *spot market*—and administered payments. If we start with the latter, we will see that the Spanish electricity bill includes a wide range of subsidies. In addition to the very visible ones received by renewable energy production until 2013, there is a long list of less obvious subsidies, all paid through the electricity bill: national carbon mining, nuclear waste storage, compensation for the nuclear moratorium, compensation for the transition to a liberalized market, special electricity prices for heavy industry (such as cement and steel), and, as we saw earlier, the "capacity payments" received by hydro- and thermoelectric power plants to provide, so the justification goes, stability to the system.[69] In 2016, these administered payments represented 35 percent of the electricity bill paid by ordinary Spaniards.[70]

On the other hand, by lumping together all energy production technologies, the spot market has been insensitive to production costs, providing windfall profits to hydroelectric and nuclear plants. Indeed, a strong argument can be made that the deficit is the result of *overpaying* for the electricity generated through traditional sources (mostly hydro and nuclear) rather than the result of consumers' underpayment.[71]

It is worth reiterating that the so-called liberalization process did not open the sector to new actors or liberalized consumer prices; what the government actually liberalized was the utilities' capacity to determine the shape and rate of growth of electricity generation, surrendering the state's ability to control and rationalize energy production expansion and to harmonize the conflicting interests stemming from it. Indeed, during the miracle, the oligopoly reinvested its windfall profits in combined-cycle gas power plants, creating a hypertrophied productive capacity. Then, with the crisis and the decrease in electricity demand, the combined-cycle plants became redundant, and have been largely inoperative for three quarters of the year since 2008 (see table 5.3). The financial situation of the utilities, heavily overinvested in underutilized capacity, quickly degraded. The oligopoly, then, needed to increase electricity prices to cover financial losses. The regulatory system, thus, created a market that was not only insensitive to production costs, but also to demand, giving rise to the paradoxical situation that decreasing demand and excess supply translated into rising prices for ordinary consumers.

While electricity demand was growing and the miracle put a premium on new plant construction, the oligopoly played the renewable game. And it kept doing so between 2008 and 2011, taking advantage of the safety of the FIT system, but also of the credit restrictions affecting independent wind developers to take over their business, thus strengthening its grip on the sector (see chapter 6). However, with the crisis, a new zero-sum situation ultimately emerged: every new megawatt of wind energy was a megawatt that the utilities lost; the expansion of wind energy worked to the oligopoly's detriment. The only way to stop the development of wind energy was to change the regulations that governed it.

As Mario Sánchez-Herrero sums up: "[The oligopoly] has had to sacrifice their only solid bet in renewables: wind energy.... Their [financial] position is so compromised that, using an analogy from the game of chess, they had to sacrifice a bishop.... They are prepared to slow down its development in Spain because the money they make with it is money they lose elsewhere."[72]

The government's reforms aimed to guarantee the oligopoly's profits and restore its financial standing, meaning these reforms—together with the international corporate lawsuits and the privileged financial aid that Spanish utilities have received from the European Central Bank—represent a massive transfer of public funds to corporate hands.[73]

The boom-and-bust cycle described by Spanish wind energy is to a large extent a consequence of the wind bubble. Playing with the double meaning of the word *contain*—"to have within" and "to hold back"—this idea can be expressed by saying that *the boom contained the bust*.[74] Indeed, the circumstances that provoked the sector's unraveling were part and parcel of the wind bubble: dominance of the utilities, reliance on a growing electric demand, unplanned expansion of productive capacity, subordination to the construction craze, liberalization of the electric sector. The wind bubble was part of a broader bubble that, while growing, impeded its bust, until the larger bubble burst. In that moment, the utilities leveraged their control over the wind sector and their proximity to political power (both central characteristics of the wind bubble) to organize the bust of wind energy.

Energy Futures

Big electrical corporations have been able to mobilize debt to be in a position of control over the system. If, with liberalization, the state surrendered its capacity to plan the development of the energy system, the crisis is pushing the situation even further. Electrical debt has developed a semiautonomous life, to the point of greatly impairing the ability of the government to set the terms of what is, in essence, a public service. The elimination of the feed-in system paves the way for big electricity corporations to increase their control of a renewable energy sector that had been growing beyond their expectations and partially out of their control. New regulation is not meant to paralyze renewable energy development in the long run; the oligopoly knows this is not possible. Rather, it is aimed at ensuring that the oligopoly can shape this development and set its timing. Framing the issue in terms of debt and invoking notions of financial sustainability, the oligopoly has been broadly successful in arguing that Spanish consumers are in debt to them, a debt that projects into the future: not only future electricity bills, but the whole energy future is owed to them.[75]

Debt is indeed being mobilized to curtail the construction of new energy futures. The oligopoly now dictates the terms—the pace, direction, and ownership—of any process of energy transition. The best example is found in a 2015 decree popularly known as the "sun tax."[76] In practice, this decree stops just short of outlawing electricity self-provision, a mechanism that many see as the best instrument not only to build a decentralized energy system, but also to help fight against growing rates of energy poverty. The decree imposes complicated administrative procedures and significant taxes on electricity production for self-provision, such as a solar panel on the roof of one's home or a wind turbine nearby, as well as heavy fines and even the possibility of criminal prosecution for those who do not comply with the regulations. The rationale is that consumers have a debt to the oligopoly, so all flows of money and energy must go through them.[77]

The sun tax reveals the main goal of the sector's restructuring: the maintenance of a centralized and concentrated electricity system. This objective was stated baldly by a government official, who was quoted in the press saying, "The problem with solar energy is that some people thought that anyone could produce electricity."[78]

Even further, in 2016, the Spanish Constitutional Court suspended the Catalan law of energy poverty, which declared it illegal for electricity companies to discontinue electricity supply to those consumers who did not pay their bills. The Court argued that the law "threatened the revenue of electric utilities and the unity of the market."[79]

Despite its undeniable positive environmental aspects, wind energy developed as part of the dynamic of growth of the Second Spanish Miracle, becoming entrenched in its relations of production and economic rationality. In an act of apparent political pragmatism, it was supported by the environmental mainstream, although the process blunted the movement's transformative potential, whereas wind energy environmental benefits were reduced as it was held hostage to the interests of the electrical oligopoly and the Second Miracle's construction-led growth. Replicating the extractive approach that has historically characterized the electricity sector, wind experienced a rush that condemned it to suffer a bust. The outcome could hardly be more onerous: not only has renewable energy development been stopped, but the oligopoly has consolidated its power and gained further control over the future of Spain's energy system. If growth became a trap, the unfolding of the economic crisis shows that the consequences of *unplanned degrowth* can also be dire.[80] The crisis has muddied any vision of a future of democratic, *planned degrowth*. All this notwithstanding, the construction of a new energy path is still possible, "It did not have to be this way," as many activists told me. Our energy future was always going to be the outcome of ongoing social struggles rather than technological fixes.

Notes

1. On the boom-and-bust cycles of guano and shale gas, see Clark and Foster (2012) and Christopherson and Rightor (2012).

2. Kropotkin (1985 [1901]), Howard (1965 [1902]), Geddes (1968 [1915]), and Mumford (2010 [1934]).

3. Mumford (2010: 161 and 158).

4. Mumford (2010: 178).

5. Mumford (2010: 158).

6. Debeir et al. (1991: 230).

7. Lovins (1977); see also Ariza-Montobbio (2013).

8. The equation between evolution and increased energy usage is explicit in classic social analyses of energy, such as White (1943) and Cottrell (1955). For a critique, see Nader and Beckerman (1978).

9. Alanne and Saari (2006: 539); see also Scheer (2006).

10. Oceransky (2010: 505).

11. Hinkelbein (2010).

12. Hvelplund (2014).

13. Szarka (2007). The five countries are: Denmark, Germany, Spain, France, and the Netherlands.

14. Lovins et al. (1999).

15. Hajer (1995: 32).

16. Redclift (2005).

17. Zizek (2006: 38).

18. Two examples of this view are Pasqualetti (2011) and Acheson and Acheson (2016); for a rebuttal, see Wolsink (2007).

19. The relationship between centralized models of wind energy and popular opposition has been highlighted in specific case studies, such as Zografos and Saladié (2012) in Spain, Powell and Long (2010) in the United States, Howe (2014) in Mexico, and Argenti and Knight (2015) in Greece, and in comparative analyses, such as Toke (2002), for the UK and Denmark, and Nadai et al. (2010) for France, Germany, and Portugal.

20. Abramsky (2010).

21. My knowledge of the history of Tecnolia is based on formal interviews with several members and former members of the company. Brief descriptions of this history can also be found in Puig i Boix (2009 and 2014).

22. The labor conflict would be settled in 2016, with a downsizing that affected two hundred thirty-six workers, eighty-nine of them in the Barcelona headquarters ("La dirección de GE-Alstom y los sindicatos pactan un ERE para 236 trabajadores," *Público*, June 8, 2016).

23. A necessarily incomplete list of these thinkers, based on my interviews with antinuclear renewable energy activists, should include the works of Amory Lovins, E. F. Schumacher, André Gorz, and Ivan Ilich, as well as the Meadows Report and the British journal *Undercurrents*.

24. TARA (1980: 113–114); these conclusions were written by Cipriano Marín.

25. TARA (1977: 118).

26. The bibliography on Danish wind energy development in English alone is vast. For a detailed overview, see Maegaard et al. (2013: 33–386).

27. Hvelplund (2014: 81). In the last two decades, this model has been eroded with the extension of larger wind farms and individual ownership.

28. This wide range is a function of technological changes and legislative parameters. In the late 1990s, turbines had a power capacity of 0.5–1 MWh; nowadays, the industry standard is 3 MWh. This has not involved a growth in wind farm size, given that Spanish legislation de facto restricts their size to fifty MWh; most wind farms are in the twenty- to fifty-MWh range.

29. On Mondragón, see Kasmir (1996).

30. See "GE concludes Alstom purchase," *Wind Power Monthly*, November 2, 2015.

31. Analyzing global data for the period 1960–2009, York (2012) calculated an actual ratio of 10:1, that is, every KW of nonfossil electricity replaced 0.1 KW of fossil electricity.

32. Martínez (2014: 240).

33. Martínez (2014: 240).

34. Wind power accounts for less than 5 percent of EU-27 electricity supplies (Barcelona 2012). For a comparison with non-European countries, see Koster and Anderies (2013).

35. GDP data from Asociación de Productores de Energías Renovables, last modified June 1, 2012, accessed October 3, 2017, http://www.appa.es/descargas/Informe_2012_Web.pdf. Employment in renewables reached its peak in 2009 with 88,050 jobs; approximately a third of these jobs were in wind energy (Garí Ramos 2014).

36. It should nonetheless be noted that, as Isenhour and Feng (2016) argue, GHG emissions are far from a perfect indicator of fossil fuel use, for they do not take into account the displaced character of much global consumption. In the Spanish case, the central role of construction related activities, and their localized character, partly offsets that bias.

37. By 2011 (i.e., prior to the elimination of production subsidies to renewables), jobs in the sector had already decreased from its peak in 2009 by more than 50 percent—from 88,050 to 55,861 (Garí Ramos 2014).

38. Jochimsen (2013: 484).

39. Charnock et al. (2011: 7).

40. Maruja Torres, "Qué pestazo!" *Eldiario.es*, October 7, 2015.

41. As Regueiro Ferreira (2011) shows, this percentage becomes higher when we take into account the financial involvement of large corporations in projects led by independent developers.

42. In 2004, Spain shifted from a classic FIT system to what is called a "floor/cap premium." Through this variant of the FIT system, wind energy producers had to sell their electricity at market prices (with a floor), receiving, in addition, a fixed premium (with a cap). Preferential access to the grid was maintained.

43. See Mitchell (2010: especially 178–197) for a comparison of these policy frameworks in the EU.

44. Regueiro Ferreira (2011) and Saladié (2013). Profits are, however, highly variable, for they are a function of the hours a wind farm can be in operation, which fundamentally depends on its location.

45. Barcelona (2012).

46. Mitchell (2010: 188); see also Szarka (2007).

47. See Gallego and Victoria (2012).

48. See "Hacia una mayor liberalización del sector electric," *El País*, December 8, 1998.

49. Carreras and Tafunell (2010: 426–434).

50. González et al. (2010).

51. For a nonexhaustive list of former politicians working for electricity corporations, see "43 políticos 'enchufados' en eléctricas," *El Mundo*, February 23, 2014.

52. Iberdrola provides the best evidence of this process. This company is the global leader in wind energy production, and Qatar's sovereign wealth fund is its main stockholder.

53. See "Hacienda detecta comisiones en la concesión de parques eólicos," *El País*, April 20, 2015.

54. See, for instance, "Oriol Pujol vetó informes de medio ambiente para lucrar a altos cargos de CDC," *Elconfidencial.cat*, August 6, 2014.

55. Harvey (1999).

56. For a general overview of these mechanisms, see López and Rodríguez (2011), Palomera (2014), and Narotzky (2016b).

57. Fernández Durán (2006: 30).

58. Carreras and Tafunell (2010: 473).

59. López and Rodríguez (2010: 336).

60. Prieto and Hall (2013: 27).

61. Data obtained through www.minetad.gob.es. As for absolute figures, total installed capacity grew from fifty-seven to ninety-five GWh. Half of this increase corresponded to natural gas power plants, which grew from seven to twenty-four GWh.

62. Prieto and Hall (2013: 118).

63. See Franquesa (2016).

64. ACA (2016).

65. For a clear expression of these efforts, see the joint press release issued by nine European electricity companies in October 2013, last edited October 1, 2017, last accessed October 3, 2017, http://www.eon.com/en/media/news/press-releases/2013/9/10/heads-of-nine -leading-european-energy-companies-propose-concrete.html. For a clear expression of the neoliberal character of EU energy policy, see European Commission (2011).

66. For discussion and analysis of this debt, see Fabra Portela et al. (2012), Matea Rosa (2013), and Franquesa (2014).

67. In 2008, the government learned that the oligopoly had overcharged its customers by 3 billion euros under this scheme; it inexplicably decided not to ask for the money back (see "Industria redactó una orden para recobrar 3,000 millones a las eléctricas," *El País*, December 8, 2014).

68. On the avalanche of cases brought to the ICSID (International Center for Settlement of Investment Disputes), see, for instance, "El gigante francés EDF se suma a la avalancha de denuncias contra España por las renovables," *Eldiario.es*, July 1, 2017).

69. Consumers' payment of the nuclear moratorium was liquidated on October 26, 2015.

70. "Claves para entender por qué sube la factura de la luz," *Público*, January 19, 2017.

71. See Fabra Utray (2014).

72. Sánchez-Herrero (2014: 100).

73. See, for instance, "Spain's utilities to benefit most from ECB debt purchasing programme," last edited June 8, 2016, last accessed October 3, 2017, http://www.thecorner.eu; see also, "El rescate encubierto de las elécricas," *Diagonal*, October 10, 2016.

74. On the meanings of "contain," see Dupuy (1999).

75. On the moral overtones of debt, see Graeber (2011).

76. It was passed on October 9, 2015, with the convoluted title of "Royal Decree 900/2015, regulating the administrative, technical and economic conditions of the modalities of electricity supply for self-consumption and production with self-consumption."

77. Iberdrola's US subsidiary made this point explicit in a recent report to the Security and Exchange Commission, arguing that net metering, while good for the environment, was a risk for the company's bottom line (see especially p34, at www.sec.gov/Archives/edgar /data/1634997/000156459017004016/agr-10k_20161231.htm#ITEM_1A).

78. See "Un Real Decreto contra el abastecimiento energético," *Eldiario.es*, October 11, 2015.

79. "El Constitucional anula el restrictivo decreto catalán contra la pobreza energética", *El Periódico*, April 8, 2016.

80. On *degrowth* and the crisis, see Schneider et al. (2010); see also Latouche (2009).

6   Accessing Wind

Man's power over nature turns out to be a power exercised by some men over other men with Nature as its instrument.

—C. S. Lewis, *The Abolition of Man*

THIS CHAPTER DESCRIBES the decade-long process that culminated in the operation of three wind farms near the village of Fatarella. When wind companies first prospected in the area, most villagers wanted the wind farms. Today, opinions are more divided, and many locals bitterly complain that wind companies "have become the landlords [*amos*] of the territory."

Most seem to agree that the development process did not go as it should have. "We wanted them, but not like this. It hasn't been done the right way. We have not known how to negotiate, or we haven't been able to," my neighbor in Fatarella, Pere, told me.

The creation of a wind farm may be described as a collection of operations oriented to secure *access* to a series of conditions and means of production: capital, labor, land, technology, markets, the electric grid, and governmental permits. By "access" I mean, following Ribot and Peluso, the *ability* to derive benefits from resources: "[B]y focusing on ability, rather than rights ... this formulation brings attention to a wider range of social relationships that can constrain or enable people to benefit from resources without focusing on property relations alone."[1] As we will see, the developers' *ability* to access wind (and profits)—his power over nature—is premised on an unequal balance of forces—power over people—that finds expression in the locals' *inability* to negotiate, their incapacity to set the value of their land and assert their worth.

The analysis of access informs us about the social relations of production structuring wind energy development in Spain, but more importantly, it informs us about the field of power and value relations within which processes of production operate. Eric Wolf calls this field *structural power*: "[Structural power] specifies the distribution and direction of energy flows ... [It is] power that structures the political economy. [It] shapes the social field of action in such a way as to render some kinds of behavior possible, while making others less possible or impossible."[2] Structural power organizes the social division of nature and labor, articulating the relationship between them.

From a formal standpoint, I describe the process that allowed the developer to access Fatarella's wind (and to turn that access into profits) as a symphony in four movements: a long, complex process involving variegated themes and multiple subplots that nonetheless possesses a thorough structure and orientation. The first movement, roughly from 1999 to 2003, describes the establishment of an agreement between the municipality and the developer that gave the latter leverage over the municipality, which he used to subject the local government to his interests. The second movement, from 2003 to 2007, deals with the construction of the electric line to connect the future wind farms with the electric grid. The third movement, from 2007 to 2009, describes how the developer accessed the land and reveals how existing social tensions and the developer's divisive strategies abruptly increased social conflict within the village. In the fourth movement, in 2009, the developer converted his access to wind into value by selling off the permits for the wind farms to a large multinational.[3]

Prelude: On Seamless Machines and Wretched Appendages

George Caffentzis observes that the grand industrial exhibitions of the nineteenth century were not only occasions for intercapitalist exchange, but also public exhibits of the power and prowess of industrial capitalism: "armed parades ... display[ing] the power of one's weapons. [Their] success was such that the machine and its power had become the literary expression of capital in general."[4] In a commercial that aired frequently during my fieldwork, aerial images of wind farms were shown cutting through at daring angles, while single words flashed on the screen: "sustainable," "competitive," "fair," closing with, "Iberdrola, the worldwide leader in wind energy production and one of the five biggest electricity companies in the world." Wind turbines have become a preferred icon of what Alf Hornborg calls the "power of the machine" and, by extension, of "capital in general."[5] Yet the image is less evocative of a military parade than a peaceful uprising.

The wind turbine represents a new kind of Spanish capitalism: green and efficient, kind and daring, simple and smart. Contrast the smoke, coal, darkness, crowds, and strikes of the Industrial Age with a three-bladed turbine: white, sleek, smooth, seamless, silent, and elegant—the emblem of intelligent design—atop a mantle of green grass, perpetual motion against a blue sky. This is wind energy's form, a platonic idea: it exists in our heads, reality never quite achieving that perfection—after all, as Roland Barthes pointed out, only sci-fi spaceships are seamless.[6] Certainly, we are captivated by the turbine's promise of renewable energy, but it is not only that: our awe is also linked to its seemingly perpetual motion. History is full of men chasing impossible machines to make something out of nothing—alchemy. From Galileo to the architects of thermodynamics, scientists demonstrated that these attempts to "cheat nature" were futile: work cannot be done without a source of energy.[7] Wind engineers and developers thus have no intention of cheating nature, but of harnessing its hidden, inexhaustible potential. Eudald, the original developer of the three wind farms currently located in Fatarella, explains it thus: "As an engineer, I've always been fascinated: how you can make something useful out of something that you don't need to pay anyone for. Because "Our Lord"—to put it in some way—does not get paid for the fluid [wind] that you use for your electric turbine, and in contrast society reaps a great benefit from it: cheap energy. The hydrocarbon cycle that we are part of needs to be closed off. We are destroying our planet, and we have the solution in our hands."

The quote is a clear expression of what Alf Hornborg calls *machine fetishism*, for it conceals the system of unequal power relations and asymmetric material exchanges that make technology productive.[8] Notice that Eudald is thinking as a capitalist, interested in making wind into energy only insofar as he can turn it into money. He knows that he cannot make something *from* nothing, but he is excited about making something *for* nothing, creating value out of something that has none. Eudald's power should be defetishized, for the value that he is making is not the simple product of his ingenuity, but rather depends on a broad field of value relations structuring the appropriation of nature and the organization of labor. These include not only the particular characteristics of the Spanish electric system and its regulation, but also more general characteristics of capitalism, such as the profit motive, the price system, and the commodification and international division of labor and land. Eudald's virtuous conversion of wind into value also depends, crucially, on the land where those turbines are sitting, and on those who own, labor, and inhabit that land. Wind turbines may reach up into the sky to create abstract value, but they are rooted in messy ground.

Take another Iberdrola commercial. We see a man in his home, speaking to the screen as he dodges the blades of a turbine installed in his dining room: "To save with Iberdrola, do I need to have a turbine at home?"

"No," an unseen voice answers, "you just need to have a contract with Iberdrola." By telling consumers that they need not bear the burden of wind energy generation, just enjoy its benefits, the ad conceals two crucial facts. First, at the time the commercial was airing, electric utilities were actively lobbying to prevent consumers from having a turbine "at home" (a struggle that led to the establishment of legal restrictions on electric self-provision in 2015); second, the wind electricity that Iberdrola offered to its clients was produced *somewhere*: in overlooked and marginalized wind energy-generating hubs inhabited by people, such as Fatarella. Locals responded to the ad, "Well, some of us *do* have turbines at home!"

The commercial reveals two fundamental conditions that allow wind developers and electricity utilities to make their profits: the separation of production from consumption and the location of wind generation facilities on land that is cheap because its owners have little voice. The location of wind farms in peripheral regions such as Southern Catalonia is not an epiphenomenal characteristic of Spanish wind farm development based on marginalist calculations, but a central component of its political economy, premised not on the existence of cheap land, but rather on the developer's capacity to pay for land cheaply.

Marx's seminal distinction between absolute and relative surplus value tells us that machines and technological innovation pose a permanent threat to the people in the lower echelons of the workforce: "When capital enlists science in her service, the refractory hand of labor will always be taught docility," wrote Andrew Ure in 1835.[9] The threat's disciplinary quality is obvious: destitute workers must be grateful, even if capital needs them to maintain the rate of surplus value. Wind energy development in Spain also needs "wretched appendages," people in places whose value can be diminished to nothing, whose inhabitants can be taught docility by being reminded that what they have has little value. Eudald's extraction of value depended as much on the null price of wind as on the capacity to access it at a low cost. It required treating the land of Fatarella as cheap—and further devaluing it in the process—and exploiting its inhabitants' lack of economic and political resources.

Libretto: How to Make a Wind Farm

The process of "making a wind farm" involves three different phases: licensing, construction, and operation. I start with the latter two because, as we will see, it is ultimately "licensing" that is the key phase of the process.

The Easy Part: Construction and Operation

For a standard, fifty-MWh wind farm, construction typically takes a year. The process is relatively unsophisticated technologically, but intensive in terms of capital and labor. First, the terrain must be prepared, and then the turbines are installed. Preparing the terrain involves a whole host of activities, but the three basic ones are building dirt tracks (or expanding old ones) for the transportation of turbines, digging ditches for the underground connection of the turbines to one another, and constructing an electric substation to collect the energy that will ultimately be sent to the main grid.

Preparing the terrain affects, directly or indirectly, a rather large area (in the case of Fatarella, about a hundred hectares per wind farm). The work is performed by the main construction company, which is either the electricity company that owns the wind farm or the turbine manufacturer. The firm usually hires local labor for temporary, low-skilled jobs such as reconstruction of damaged properties, small earth-moving operations, or non-turbine transportation. The transportation and installation of the turbines are the most complex parts of the process, and the main construction company often subcontracts these operations to specialized firms. For instance, in Fatarella, the cementation was carried out by a crew of Portuguese workers; the turbines were transported from the northern Atlantic part of Spain—where they arrived by sea from Germany—to Terra Alta, a trip of almost a thousand kilometers, by a Spanish firm; and the on-site turbine assembly was coordinated by a British company.

This process is expensive. According to the wind farm owner in Fatarella, the project cost upward of 30 million euros, half of which was absorbed by the purchase of the turbines. The highly competitive turbine market makes access to this technology unproblematic, but the construction phase can, of course, be held up by the developer's access to capital. Nonetheless, since the feed-in tariff system guarantees predictable profits and there are few technology risks involved in the construction process, it is, at least in principle, easy for wind farm developers to raise money in capital markets.

Additionally, despite the variable quality of wind, wind farm developers measure the wind on their proposed site and have a good idea of the quantity of electricity that they can generate every year.[10] Eudald explained the steps he took to access capital: "I did a *project finance*. For this, I constituted an SPV, a special purpose vehicle, which was a company [called Swellwind] that I created ad hoc to manage the future wind farm, and I transferred all the agreements and licenses to it…. I add the permits, I take it to a bank, and I demonstrate that I have the wind, I demonstrate that I have located the appropriate machine for the production of energy, and I demonstrate that I have a regulatory framework that guarantees [prices and market access]."

Operation, too, involves minimal risk and fairly small numbers of medium- and highly skilled workers. Each company has one or more remote control centers that monitor and maximize electricity production: for example, calculating the optimal direction of the blades, detecting malfunctioning turbines, predicting future production, and so on. Each wind farm also has one or two local employees, often with technical degrees, to perform regular maintenance operations. Finally, since the market was guaranteed under the FIT system, which operated up until 2013, the sale of electricity to wholesalers was unproblematic, thus dissipating any risk of overproduction.

Access to capital, labor, technology and markets thus constitute barriers that the developer can easily overcome. Eva, an engineer working for Eudald with a great deal of experience in wind farm development all over Spain reinforced the idea that these two phases were the "easy" ones: "Licensing is hell…. Building is incidental: the important thing is all that happens before. And then there is the operation of the wind farm, which, there is no mystery to it." Barring highly improbable eventualities, such as the destruction of the turbines or retroactive changes in the regulatory framework, once the wind farm is licensed and construction starts, it is very unlikely that a wind farm project will derail.

The Hard Part: Licensing

Licensing involves gaining access to three fundamental conditions of production: the general electric grid, land, and bureaucratic permits. This phase has low capital demands and relatively low labor demands, but it is very time-consuming and involves coordinating actors with divergent interests, thus requiring large amounts of social and political capital. The careful cultivation of a mix of relations of trust and power, as well as the political contacts I discussed in chapter 5, is critical.

I should acknowledge that the term "licensing," used by developers to refer to everything that happens before the actual construction starts, is slightly confusing. First, we should be aware that we are not dealing with one but multiple permits (or "licenses") issued by different levels of the state's administrative apparatus. In this category we may include a wide variety of bureaucratic milestones faced by the project: obtaining certification as a renewable energy facility ("getting a REPE," which entitles the project to receive production subsidies); earning designation as a project of public interest (*utilidad pública*, which entitles the project to invoke eminent domain); approval by the provincial and regional (Catalan, in this case) administration, contingent on a favorable assessment of environmental impact; and gaining a construction permit granted by the municipality. The last of these—municipal approval for the location of the wind farm—is the single most decisive condition of the licensing process.

Further, a good deal of what goes on during the licensing phase has little to do with obtaining permits. Fundamentally, a developer needs to have access to the electricity grid and, most crucially, to land. While the state formally guarantees right of access to the grid to renewable energy producers, in practice, this access may be more complicated (as we will see). Access to land involves reaching agreements with the landowners, as divided into two categories based on the use of the terrain. On the one hand, owners of parcels directly affected by turbines are regulated through land leases involving a sizable annual payment. On the other hand, owners of parcels affected by related infrastructure—tracks, ditches, and electric lines—receive a one-off payment as compensation for the right of access and any damages or losses it may involve, such as the removal of crops. In both, although the impacted area is generally quite small, the contract between landowner and developer affects the whole parcel of land, so as to assure the developer that the landowner will not transform the parcel in any way that is detrimental to the developer, such as authorizing the placement of a turbine from a different developer or radically transforming the built environment.

During the licensing phase, the developer faces multiple uncertainties: not obtaining the construction permit; being denied access to the grid; being unable to reach agreements with landowners; the election of a new municipal government opposed to the wind farm project; receiving an unfavorable environmental impact assessment; and so forth. Local organized resistance is, by far, the most frequent cause of these diverse potential setbacks; it's what, in Eva's words, makes this process "hell" for the developer. "Organized resistance" encompasses a wide spectrum of strategies, including, for instance, preservationist groups filing environmental lawsuits, a local political party opposing the wind farm, landowners rejecting contracts, a platform organizing a campaign against the project, or a municipality enforcing the law with scrupulous zeal. While the opposition of the regional and Spanish governments may be fatal for a project, such opposition may only emerge as a result of local resistance.

Thus, the licensing phase may be viewed as a process of bypassing local resistance. To this end, the developer makes use of a series of mechanisms and strategies: expropriation, public relations campaigns, paying attractive compensation to a critical number of landowners, leveraging a position of control over the municipality, building on internal divisions in the community, and so forth. Indeed, given that licensing is the choke point of wind energy development—and by extension the whole wind energy industry in Spain—the ability to prevent the emergence of, and ultimately overcome, local opposition is the crucial element in accessing wind.

First Movement: The Agreement

The origin of the three wind farms that currently surround Fatarella lies in the encounter between the independent municipal government of Fatarella—backed by Unió de Pagesos (UP) and historically aligned with antinuclear positions—and Eudald, the heir to a prosperous, mid-sized Catalan construction firm.

"Do Not Sign!": Controlling Access

In Terra Alta, in the late 1990s, Enron tried to harness the potential tension between landowners and municipalities, which must authorize the development of wind farms, by offering private contracts to landowners without negotiating with municipalities. The company believed that signed landowner contracts would force the municipalities to come to agreements with it. Having tried and failed to create a joint strategy with the other villages in the county (see chapter 4), the municipality of Fatarella focused its efforts on building a more united approach within the village. The reaction from the municipality to Enron's efforts was quick and clear: "Do not sign! Wait." This message was announced through the loudspeakers that twice daily inform the villagers of news, from the price of fish at the local fishmongers to meetings of the school parents' association.

Bernat, a municipal councilman between 1999 and 2003, explains the rationale of the announcement: "We needed to stop that: no one moves! Because, by signing, they would have allowed the company to be the *amo* of the territory. Some did sign, but very few, so we did stop it. Because we wanted to oblige the company to make an agreement with the municipality, establishing some principles, so that we could negotiate in the right condition. Because how can you negotiate if a big group of people have already been promised money and they expect it? Your hands are already tied."

Fatarella successfully barred Enron's access to the village's lands, while Corbera and Vilalba cut deals with the conglomerate. Over the next few years, the municipal government of Fatarella had conversations with at least four developers, and finally settled on Obrisa: "We thought it was more serious and smaller, and we also liked the fact that it was a Catalan company. Did we make a mistake? I don't know," muses Bernat.

Obrisa: In Search of Sustainability

Obrisa was a family-owned construction firm with a typical Spanish trajectory.[11] Founded by Eudald's grandfather in the 1950s as a small earth-moving company, it benefited from the constructive expansion of the First Miracle. By the 1980s, the firm was a sizeable outfit playing in the lucrative field of large-scale projects, successfully bidding for the construction of new airport terminals, highways, irrigation projects, convention centers, and hydroelectric projects. During the Second Miracle, Obrisa created a holding company that facilitated its access to new financial resources that allowed it to open branches in Chile and Colombia. It also entered the Spanish wind energy sector by initiating the procedures to receive authorization to build several wind farms in Catalonia (four in Terra Alta and others in Empordà). Eudald describes the move into wind energy as a personal bet and his dearest project. Aware that the construction bubble would come to an end sooner or later, he reckoned that wind energy would provide a reliable source of income that would *sustain* the continuity of his family business.

Eudald is quite proud of his wind farm licensing record, which he explains as the result of two different qualities. One is his knowledge of the "mentality of the *pagès*:" "The big corporations go there with a trolley suitcase and start speaking about *business*, and they visit the mayor, who arrives from the fields on a tractor.... The locals don't trust them. But with companies like ours, it is different. We speak the [Catalan] language, we don't wear a tie, and we understand the mentality of the *pagès*."

As we have seen, the progressive municipal government of Fatarella certainly valued the fact that Obrisa was relatively small and local, implicitly comparing them with the oligopolistic electric utilities. On the other hand, Eudald's capacity to license wind farms also clearly stemmed from a deep knowledge of the Catalan administration, achieved through the cultivation of political links for half a century: "We got the permits because we know the local and regional administrations, and we have a good image of being serious people, so *bam, bam, bam* [we got it done]." Indeed, over the past few decades Obrisa has been one of the main beneficiaries of public contracts, and it is widely considered to be very close to the Catalan government and to CiU, the ruling party between 1980 and 2003.[12]

The Referendum: Preventing Division

By 2000, negotiations between Obrisa and the municipal government were advanced. Nonetheless, the municipality was aware that this was a potentially divisive issue, especially given the political context in the region, with rising opposition to wind farm development and the beginnings of the Southern Revolt. Fèlix, who was the mayor at the time and a past leader of UP, explains: "We feared that this issue of the *molins* [turbines; lit. mills] could erode the *convivència* [cohesion] in the village, so we did a series of things to inform the population and we made everyone be part of the decision." The term *convivència* is important: it could be translated as "coexistence," but it emphasizes the bonds articulating a certain collectivity, expressing the positive qualities of "living together." We should interpret the bulk of the municipality's decisions in this early period as guided by a desire to foster togetherness and prevent division.

In 2001, the municipality organized three town hall meetings to debate the issue, inviting speakers with diverging positions: a representative of the Platform of Terra Alta, a representative of the energy branch of the Catalan government, and Xevi, one of the founders of Tecnolia, in his position as both environmentalist and industry insider (see chapter 5). This process culminated in a referendum (celebrated despite Obrisa's reluctance), a feat that can hardly be overemphasized: no other village in Spain had ever organized a popular referendum to make a decision about wind farm development. Despite low voter turnout (around 30%), the result was clear: two hundred twenty-eight in favor of the installation of wind farms in the village versus one hundred five against. The referendum asked two additional questions of those voting in favor, concerning the number of wind farms that the village could host and their location. A majority said that they wanted just one wind farm and that it had to be located at a "prudent distance" from the village nucleus. The municipal government had its mandate: to create the conditions for limited development of wind energy in Fatarella.

This result reflects both the desire to find a reasonable compromise between the different sensibilities in the village and the limited enthusiasm that wind energy development elicited even among its supporters. As a general rule, only two kinds of landowners were unconditionally supportive: retired farmers with no offspring in the village and those who, as a result of emigration, did not live there. Opposition was especially common among those who could count on good, nonagricultural salaries (for instance, from the nuclear power plants) and those who practiced agriculture as what they describe as "a hobby, just to maintain the land." The large number of semi-proletarianized villagers for whom agriculture was a substantial, but not sole source of income saw the wind farm as a potential source of revenue to complement their limited incomes.

Bernat and Pere, neighbors and cousins with opposed political ideas and little personal relationship, fit this latter description. Both are in their late forties and have a good amount of land, working as full-time farmers, but neither can live off the land alone. Their households rely on their temporary jobs or side businesses and their wives' part-time work as supermarket cashiers. Their stance toward the wind farm may be described as cautious pragmatism: "Everyone likes having his fields clean [with no turbines], but I felt that this was an opportunity that we couldn't say no to, a resource that we had to take advantage of," Pere told me.

Note that he supported the wind farm because he couldn't say no, in other words, because he cannot live off the land alone. In this respect, the mildness of his and others' support stemmed from the fact that the wind farm, as they often said, "doesn't fix anything," meaning that it would not solve the semi-proletarianization and precarious livelihoods of the locals.

What the inhabitants of Fatarella wanted were more jobs and higher prices for their agricultural produce. The wind farms did not offer any of that. Bernat expands: "We had to take advantage of it, a few thousand Euros every year are sweet, it is like extending your harvest. But let's be clear, they don't fix anything.... I always said it: we take advantage of a resource that we have, fine, but this doesn't give us jobs, this does not solve our problems of depopulation and all that, we just get a lease, and we need to live here. We are not here to rent terrains, we have to live here, and for that we need to be able to work."

The Agreement: Obliging Loyalty and Securing Access to Land

After years of negotiation, in 2003 the municipality and Obrisa signed an agreement (henceforth the Agreement) regulating the conditions for the installation of the wind farm.[13] This thirteen-page document, initially valid for four years, covered two main areas: the economic compensations to be paid by Obrisa, and the conditions governing the relationship between the developer and the municipality.

When it comes to the payments, my interviews and documents related to the negotiations suggest that the municipality and the developer each had a major concern. The municipality, prioritizing the preservation of *convivència*, wanted to find formulas to spread the economic benefits of the wind farm to the population at large. Thus, it demanded a fair distribution of land leases, requesting that the total amount for individual turbines be divided into two halves: one for the owner of the terrain, and the other one distributed among neighboring landowners within a perimeter of one hundred and fifty meters. The municipality also demanded that an annual payment of ninety thousand euros be made to it. In the Agreement's preamble, this payment was justified in the following terms: "energy generation ... must adequately compensate the inconveniences that producing facilities cause to the affected municipalities and the people who reside there, for [these facilities] satisfy a demand that is located in other territories."

Interestingly, the Agreement referred to this payment as a *cànon*, a term borrowed from the economic compensation that the municipality receives from the state for being located near a nuclear plant. These details reveal that the municipal claim to perceive an economic compensation was rooted in an oppositional attitude, especially vivid in the context of the Southern Revolt and informed by a long local history of antinuclear struggle and distrust toward energy companies. Indeed, although these kinds of municipal compensations have become quite common throughout Spain, the *cànon* was a typically Southern Catalan invention of the early 2000s. The oppositional origin of the *cànon*, and even its mere existence, baffled developers such as Eva: "What is this? I've been developing wind farms all over Spain and never heard of it. We just paid the land leases ... but *cànon*, this quantity that you pay just because? I had never seen it. If I go to a village and open a factory I don't pay, quite the contrary in fact!"

For his part, the developer was worried about the difficulty of individually negotiating contracts with a very large number of landowners. Indeed, as early as 1998, Obrisa sent a fax to the municipality expressing unease about the challenges associated with the fragmented landownership structure of the village. This problem was largely solved in the Agreement, for it stipulated fixed quantities for land leases (sixty-eight hundred euros a year per turbine, divided into the aforementioned two halves), to be paid only once the wind farm entered into operation.

This arrangement largely worked in favor of the developer. Indeed, although the installation of the wind farm on particular land parcels was contingent on the signature of individual contracts with landowners, the economic terms had already been set by the Agreement, leaving the owner with only two choices: either accept or reject the deal. Yet the landowners of Fatarella frequently complained that even that binary choice had been pushed out of their reach, a complaint voiced through a recurrent expression: "It doesn't matter whether I accept or reject the land lease, because they'll move the turbine to the neighbor and they'll build the farm anyway." This expression points directly to the municipality's vested interest in the project, a point that, as we will now see, was made explicit in the Agreement.

Indeed, if we move to the general conditions regulating the relationship between the developer and the municipality—and by extension, between the wind farm and the villagers—we may observe two different sets of clauses extending reciprocal obligations. On the one hand, the municipality set a series of conditions for the construction of the wind farm: hiring local labor, respecting vernacular landscaping and local archaeological heritage, and fomenting the development of local tourism by opening the wind farm to visitors. Yet these agreements were phrased in very generic, often conditional terms, and the document did not include enforcement measures. As a result, they were poorly implemented: little local labor was hired and no visitor's center was ever built. Local architecture and archaeological elements were generally respected, but largely due to the permanent monitoring by landowners and some municipal authorities.

On the other hand, the developer introduced a set of clauses that granted monopoly rights to Obrisa, such as forbidding the municipality and landowners to negotiate with other developers, and allowing the developer to sell the wind farm without consultation. Most importantly, the Agreement committed the municipality to collaborate with the developer, aiding Obrisa in its negotiations with the public administration to obtain permits *and* with the local landowners to sign private contracts, "preventing and limiting possible conflicts ... [by] assuming a mediating role in the relations with affected landowners."

The Agreement established the basis of a relationship of obligation between the municipality and the developer that enrolled the municipal government in the developer's project and progressively blurred the lines between the two. This relationship of collaboration was, however, deeply asymmetrical. Whereas the municipality had little force and fewer mechanisms to oblige the developer, the latter was able to oblige the municipality to collaborate. The main mechanism through which Obrisa achieved this goal was a series of advanced payments on account of future *cànons* and taxes.[14] These annual advance payments represented a nice addition to the budget of a small, poor municipality such as Fatarella, which, at the time, was renovating and expanding a senior health care center. Yet they also created a strong obligation and the threat of future debt. Indeed, prior to the signature of the Agreement and unbeknownst to the public, the juridical services of the province had recommended that the municipality refuse these advance payments, which could impose an unbearable financial burden in the event of the project being delayed. The developer would be able to mobilize this framework to ensure the collaboration of the municipality, de facto a sine qua non condition for obtaining permits and for the success of the negotiation with landowners.

The neighboring municipalities of Corbera and Vilalba soon adopted the same terms with Obrisa. Indeed, during the negotiations over Fatarella's Agreement, Eudald took advantage of Enron's demise to extend Obrisa's influence over the region's wind, assuming—and later modifying—Enron's two wind farm projects in Vilalba and Corbera. These two wind farms were to be located in land that administratively belonged to those two municipalities, yet was quite far, and actually not visible from, their residential nucleus. The power plants instead were to be sited in stretches of agricultural land largely surrounding Fatarella and owned by its inhabitants. In 2005, Eudald, making use of the capacities granted by the Agreement, transferred the property of the three wind farms—Fatarella,

Vilalba, and Corbera—to Swellwind, a subsidiary of Obrisa created in 2001 that, at the moment of the transfer, had capital of four thousand euros.[15]

Second Movement: Accessing the Grid, Defeating Resistance

All wind farms need access to the long-distance electric grid. Spanish legislation on renewable energy obliges the operator of the grid—the state-controlled Red Eléctrica Española (REE)—to provide a point of connection to all wind farms. It is up to the wind farm developers to build their electric line to that point. Eudald and the rest of the owners of wind farm projects in Terra Alta asked to connect their installations to a four hundred-volt electric line crossing Terra Alta through Fatarella. REE denied that request on grounds of "national security," arguing that the variability of wind energy could introduce instabilities to that electric line, which connects Ascó's nuclear plants to a large thermal power plant situated in Escatrón, about a hundred kilometers up the river Ebro. The developer appealed this decision to the National Energy Commission, the state's energy regulator, forcing REE to provide a new point of connection in the hydroelectric dam of Riba-roja, beyond the northern confines of the county, in the fall of 2003.

Developers were deeply disappointed, for the new, more distant point of connection grossly added to their costs. The Catalan government was also unhappy. The Spanish government had established renewable energy production targets for each *comunidad autónoma*, and Catalonia was far from meeting them, in large part as a consequence of the Southern Revolt and widespread local opposition. Now, REE's decision was imperiling the main cluster of wind farm projects in Catalonia. In reaction, the energy branch of the Catalan government promoted the cooperation of the different developers in the county, brokering the creation of a private consortium called Terraca—"it was a matter of industrial planning," a government official told me. In 2005, Eudald became Terraca's first CEO.[16]

Terraca was charged with building a 24.5-million, forty-kilometer-long electric line traversing Terra Alta from south to north. This line would connect ten wind farms from eight different villages to the general grid, with each company responsible for building the part of the line corresponding to its project. The scope of the project was considerable: just the part of the electric line that was linked to Obrisa's wind farms affected more than a hundred landowners. When concluded, the ten wind farms included in the formation of Terraca would have a total installed capacity of four hundred twenty-four MWh, involving a total investment of 480 million euros, distributed among one hundred eighty-three turbines affecting a total land area of one thousand hectares, all to cover 2 percent of the total electric production of Catalonia.[17] To date, eight out of the ten wind farms have been built.[18]

In the summer of 2003, Eudald assembled a small team of engineers to carry out the licensing of his wind farms, including the electric line connecting them to the grid. To lead the effort, he hired Eva, an old classmate from engineering school with significant experience in the sector. To Eva's dismay, her arrival in Terra Alta coincided with REE's decision:

> Negotiating with the landowners for a wind farm is one thing, because it is flashy, [and] it brings in money, but the electric line is very different; you just put in infrastructure. That doesn't bring in money. And it was an easy target for the platforms. In a wind farm, you pay a thirty-year lease, but for the line you just pay once—but it is rural land, it is shitty land, and you never pay big quantities.... It's just nickels and dimes, because you just pay the owner so that he does not plant an olive tree underneath the cable.

As Eva suggests, Terraca's project revivified the Platform of Terra Alta, which had languished for a couple of years. Suddenly, the roads of Terra Alta were filled with graffiti, many still visible today: "Don't sign."

"Don't Sign!": Reactivating the Revolt

Marcel and Oriol, two young farmers of Corbera, were central in the Platform's revival. The two friends had decided several years earlier—when they were both twenty—to continue their families' agricultural activity. This involved purchasing additional vineyards; cultivating a combination of olive, cherry, and almond trees; and supplementing their income by working in temporary specialized agricultural wage jobs, mostly vine-pruning in Priorat.

Marcel, the more politicized of the two, took the lead in joining the platform. His politicization was directly related to the history of his village. During the war, the Francoist army bombed Corbera heavily (an image of the village's crumbling church made headlines all over Europe). The extent of the destruction was such that the inhabitants progressively abandoned their old houses, building a new village next to the old one. Like most of the children in Corbera, for whom the old village became a favorite playground, Marcel asked questions about the war, but the answers always left him dissatisfied: "They always talked about reds and fascists, you know, as if it were just about two different armies, and never a word about the republic and even less about anarchism. But one day I was listening to a punk group, and they were talking about the war and the social revolution and I said to myself, 'Right, that's what I always thought!' At that moment I decided I would join the CNT [main Spanish anarchist organization]." Later, Marcel created an anarchist youth collective in Corbera, of which Oriol was also part, that organized concerts and conferences on the war and anarchist thought.

In 2005, the two friends were disgusted to learn about Terraca's plan and the number of apparently dormant wind farm projects associated with it. One of Oriol's land parcels hosted a stretch of the electric line and an electric post. "One day I received a phone call from the municipality, asking me to go there because I had that [electric] post going to my land. They offered 2,400 Euros, and tried to buy me: 'If you say yes here and now,' they told me, 'you can walk out of here with half of it in cash.' You see, all very mafia style. Maybe if I had been fifty I would have said bring the suitcase, but I was twenty-five and just starting my life project."

This was a common experience. Many landowners complained that the cash advances smacked of corruption and that the fact that the negotiations took place in municipal buildings created confusion: "Am I talking to the developer or to the municipality?"

Marcel and Oriol's anarchist collective organized a series of conferences and debates on wind energy where they met the members of the Platform of Terra Alta. The encounter was crucial in revivifying the platform over the next few years: filing law suits, meeting with politicians, and organizing protests all over Southern Catalonia, including some direct actions such as the occupation of governmental offices. During these events, Marcel often carried a black and red anarchist flag reading: "They want to destroy our future." During this period of intense activity, the platform adopted a new slogan—"Terra Alta, land with dignity" (*Terra Alta, terra digna*)—and struggled to connect their opposition to wind farm development with the memory of the Southern Revolt (see chapter 4).

In their 2007 manifesto—titled "It's enough!"—the authors wrote: "Until recently, we put up with everything like submissive subjects.... But since 2000, with the struggle against Enron and PHN [the water transfer from the Ebro river], we gained our dignity and learned to organize and make ourselves respected. Never again resignation.... We are not submissive subjects anymore, we are citizens who have learnt how to struggle and we will struggle for our future."[19]

Despite this surge in activity, the platform suffered notable internal problems and difficulties in articulating a cohesive message and expanding its social base. Internally, there were growing disagreements between the old moderate sector, formed by professionals linked to conservationist organizations like *La Serena* and Gepec, and the younger, more radical sector of newly arrived members such as Oriol and Marcel. The former favored negotiation and compromise with political leaders; the latter rejected the installation of any wind farm in Terra Alta and favored direct action and assembly-based forms of organization. This clash led to the progressive distancing of the old guard, and the consequent erosion of the platform's social base and outreach capacity. This division, marked by the Southern Revolt's loss of steam and the arrival of new, more polarizing energy projects, made the Platform's task even more challenging, Marcel says:

Some people say that the *molins* are okay, but they are thoroughly against the nuclear cemetery [the nuclear waste storage facility projected in Ascó]. But some of us think that the *molins* and the cemetery are the same. It is not about the concrete activity, it is about why they build it and why it is always here. It is a structural problem. [In 2007], we tried to put all the struggles together again: the nuclear cemetery of Ascó, the gas storage facility of Alcanar, the Platform of Priorat and us, and we even organized a demonstration with a thousand protesters. But the problem is that the PDE [Platform in Defense of the Ebro] didn't want to join us, they just wanted to talk about water. Some people say, "What do you want, the nuclear cemetery or the wind farms?" This is not how it works, you don't choose between two shitty things. We want to assemble the struggle. The territory is what it is. If you love the land where you live and some people come with a noxious activity—and it is noxious because of all the interests behind it—and you let them do it, they are attacking your dignity and the dignity of the land where you live. It's like a labor issue: if they fire you, you say, "I am not an object that you can treat however you want," you then are defending your dignity, you don't accept those aggressions.

Despite its failure to mobilize the county's population, the platform succeeded in creating a climate of mistrust toward wind energy companies. Thus, although a poll commissioned by Terraca concluded that about two thirds of the county residents were favorable to wind energy development, the fact was that the landowners affected by the electric line showed very little interest in signing the contracts that the consortium offered to them, an apparent contradiction that suggests that the support to wind energy was largely in the abstract.[20] The landowners' reluctance to sign the contracts was causing a severe delay on the wind farm projects, originally planned to be up and running by 2007.

Comparing the wind farms with past projects, some landowners complained that the economic compensation Terraca was offering was too low. However, money does not seem to have been the main cause for the widespread rejection of land agreements. A better explanation must take into account the private nature of Terraca and the fact that the locals did not feel invested in a project from which they did not expect to derive much direct benefits.

Eugènia's experience confirms this idea. She headed a team of engineers specializing in agrarian land assessment hired by Eudald to assess the compensation for Terraca's line (and, later, his own wind farms), while, in parallel, also being involved with the signing of land agreements related to a publicly funded irrigation project in the county. Despite the formal and legal similarities—both one-off payments for the right of access to land—the landowners' response to the agreements was antithetical:

With the irrigation project I paid very little, like, for a pipe, twelve cents a square meter. For an electric cable, I was paying 1.4 Euros, again per square meter. Well, with the irrigation I signed ninety-nine percent of the agreements; with the wind farm stuff, instead, I had to make a lot of effort ... and we maybe reached fifty or sixty percent of agreements. With the irrigation I signed forty agreements in a day; with the electric line it was about going one day to the home, and going back again, and trying to make amendments to do it this way, and move the thing a bit to the side, and I buy you three olive trees, 200 Euros each, even if I won't touch them but just in case.... And you know why? Because the irrigation project is for them, and the other thing is, in their eyes, for people from Barcelona that "come here to steal our money," who thanks to me will make money. There's a lot of envy, and the environmentalists and the platforms have misinformed [people] a lot.

Irrigation was a publicly funded project to meet a local demand; wind development would benefit city dwellers and developers. Further, the construction of Terraca's line was making the local population aware of the side effects of wind farm construction (substations, electric lines, etc.), and this was rapidly eroding their tepid support for wind energy development.

PR Counteroffensive

Holdups in land agreements were severely compromising Terraca's line and, by extension, the whole development of wind energy in Terra Alta. In response, Terraca undertook a public relations campaign with two components. First, Terraca hired a lobbying firm. Andrés, the head of this firm, describes the local attitude toward the wind farms in the following (probably hyperbolic) terms: "Locals did not want the *molins*.... They didn't want them and they didn't know what they were about and they had absolutely no interest in them."[21] In 2005 and 2006, Andrés courted the local mayors and enrolled mainstream environmental organizations (Greenpeace, Ecologistas en Acción) in a media campaign to improve the image of wind energy development in Terra Alta.

Terraca also commissioned a team of economists to assess the future economic impact of wind energy in the county.[22] Southern Catalans have seen many of these studies: every new infrastructural project seems to come with a study disparaging the existing socioeconomic situation and emphasizing the potential benefits of the proposal. Terraca's study was no exception, presenting the county as a basket case. Agriculture was stagnant and unable to transform along "modern and rational" lines, while the fabric of small, locally owned industries—wine cellars and cooperatives, paper transformation industries, furniture, garment workshops, and so forth—was assessed approvingly, but considered insufficient to absorb the local labor supply. This critical situation would only get worse, the study said, once the construction bubble burst and the general economic situation of the country worsened. Because the study struggled to make clear that the county lacked the resources to develop endogenously, its conclusion was unsurprising: the county needed outside investment. The wind farms, with a total projected investment nearing half a billion euros, would bring that investment. Nonetheless, when it came to the details, the study had troubles to make its point, calculating that, unless additional measures were undertaken, the impact of the wind farms on GDP would be a meager 1 percent, reflecting the few permanent jobs that would be created.

The inhabitants of Terra Alta had heard unfulfilled promises of development too many times to be impressed with Terraca's line of argument. Gabriel, the young mayor of the village of Pinell, says: "When you hear certain things, you feel you are watching a TV program for kids. Always the same arguments: for the wind turbines, for the Ebro [water] transfer, the nuclear waste storage ... 'They will bring jobs, they are not polluting, they are not illegal,' (among other things because they make the law to their own measure). Dignity means not to be disrespected, not to be treated like kids. It is like the person who possesses agricultural land and prefers to ask for charity. Don't you see what you are doing?"

Gabriel is not opposed to wind energy development. In fact, he hopes that a long-planned wind farm will soon be constructed in his village. What he objects to is the idea that energy facilities can solve all of Terra Alta's problems, and the way locals' dignity is insulted by those interests that exaggerate the benefits of the projected infrastructure by disparaging the county's existing situation. For Gabriel and others, these studies treat the locals like kids who can be misled with promises of candy—an idea that finds its emblematic expression in the tale about the inhabitants of Ascó being told that the nuclear plant was a chocolate factory (see chapter 2).

Expropriation

The economic study and Andrés' lobbying had little impact on the local population. Nonetheless, Terraca's PR campaign did accomplish its main goal: in 2006, the Catalan government declared the electric line to be in the public interest (*utilidad pública*). With this designation, the state could invoke eminent domain at Terraca's request, expropriating the land it needed to build the electric line to access the general grid. Although expropriation (and threats of it) was controversial and antagonized broad sectors of the local population, it was by far the single most important mechanism that allowed the wind companies to overcome local resistance.

The designation of public interest eroded the platform's efforts to organize the affected landowners; it made resistance futile. With the designation in hand, Terraca sent a second (usually lower) offer to landowners who had rejected the first: if they did not accept it they would be expropriated. Most accepted. Yet quite a few, like Oriol, preferred expropriation to surrender (receiving seven hundred euros for half a hectare of vineyard). He says that now he does not like going to farm that parcel; it reminds him of his defeat: "My parents and grandparents had already put a fight against the Ascó-Escatrón [electric] line decades ago, and had told me that I shouldn't get into this mess, that there was nothing I could do, but I had to try it. I had to show that I cared, that I had decided to stay here and make a living here ... Otherwise, how could I tell my daughter that she should live here? If you have self-esteem, you have dignity, [and] you'll say enough."

A few months earlier, Marcel and Oriol had posted a poem on the platform's website devoted to the county. Entitled "I am wealthy," the ode describes, in rather bucolic terms, the beauty of the area, and should be understood as an attempt, central to the struggle for dignity, to assert the intrinsic value of the region and its inhabitants. It concludes by saying, "Don't steal my poverty, for money I want not."

What is most striking about the whole process of expropriation is the number of landowners who never accepted an offer. In Fatarella, there were as many as twenty-four expropriations linked to Terraca's line. Based on an analysis of the *actas de expropiación*, the judicial documents that signal the beginning of the expropriation procedure, we can identify two categories of landowners. One group consists of owners who could not be identified or located—a fairly widespread circumstance in rural areas, since in Spain there is no obligation to register rural property titles. The second group of expropriated parcels that I identified had one thing in common: their parcels contained drystone dwellings. In these cases, the rejection of Terraca's offer obeyed a cold calculation informed by past experiences.

In the *actas de expropriación* we can read the reasons landowners gave for rejecting the company's offer. Take, for instance, Mari: "The owner does not agree with the price offered by the beneficiary [Terraca], and [the offer] does not take into consideration the *perxe* [a kind of drystone dwelling] that the owner has repeatedly requested be taken into account. The beneficiary manifests that the line does not affect the *perxe* and therefore it has not been assessed.... The owner manifests that the fact that the line goes above the *perxe* [prevents it] from being modified, restored or expanded."

At first glance, this seems like a simple fight over compensation, with Mari trying to get Terraca to increase the payment. But it is not. The last line is a direct reference to the contracts Terraca was offering. In these contracts, a series of controversial clauses specified that the landowner could not introduce any alterations to the parcel of land that might jeopardize the security of the electric line. These clauses applied to the whole parcel, not just to the specific surface affected by the infrastructure. In most cases, these stipulations were plain common sense, such as not planting tall trees underneath the electric line or leaving a small, uncultivated perimeter around the electric post. Other clauses were more ambiguous, such as the prohibition on "erect[ing] any kind of building"—including temporary ones—and on "carry[ing] out any kind of action or activity that could damage or disrupt the good operation" of the projected infrastructure.

Many locals believed these clauses made the company the effective *amo* of their land: "Once you sign that thing, they are the ones who decide what you can do and what you can't do." Be that as it may, the contracts clearly involved some cession of the exclusivity rights that the inhabitants of Fatarella associate with landownership, and many viewed the drystone dwellings as particularly vulnerable, for they rarely possess any legal status. Local inhabitants hold these dwellings in high esteem and spend weekends and summer evenings in them, constantly modifying, restoring, and expanding them to the point that some have been converted into comfortable cottages at the center of local families' social life.

In the *acta* of another expropriated landowner, Marçal, we observe an effort to capture the material and sentimental value of his property:

> The landowner manifests that underneath the electric line there is a drystone *mas* of ninety square meters, 56 hundred-year-old olive trees, forty young almond trees, and twenty-one medium-sizes ones, three fig trees, three young apple trees, ten [grape] vines for eating, one louquat tree, one apricot tree, one peach tree, seven stone walls (four of sixty meters, three of seventy meters, and they all have stairs); one pool with a capacity of 50,000 liters, stocked with fish in them; two underground pipelines to feed the pool; an underground electric cable; two hoses to water the estate; five boxes of bees; and sixteen square meters cultivated with saffron. He also manifests that a tractor is continually running through the estate. He also manifests that the *mas* is his second house and has a great sentimental value, and that [the parcel] has been farmed [by his family] for more than 4 generations.

Although expropriation offered little compensation for the permanent loss of the right to alter property, it only covered the specific piece affected by the electric line. In contrast, signed agreements gave rights over the whole parcel. Thus, the landowners who, like Marçal and Mari, preferred expropriation over an agreement did so in an effort to preserve the value and control of their estates, even if that meant losing part of them.

When I first heard this reasoning in 2010, I could not help thinking that it was tinted with both naiveté and a certain amount of misplaced paranoia. Subsequent events showed that it was informed by caution and experience; the owners of the wind farms initiated judicial actions to dismantle some of the drystone dwellings belonging to landowners who had signed contracts and land leases.[23]

After Mari's land was expropriated, Terraca signaled the expropriated pieces with stakes, but as late as 2012 she told me, "They do whatever they want, the workers come and go through my fields as if it was their home." She finally was able to talk to the foreman and told him that they could not enter private property. He answered that her field was "not private property, but the forest [*eso es el monte*]."

Such incidents proliferated during the construction of Terraca's line and, later, the wind farms: landowners who were told that the expansion of a dirt track would involve the removal of half a dozen almond trees would one day discover forty trees missing; others found workers on their property even if the company had only signed a contract with the owner of a neighboring field. Still others learned that the construction of a ditch ended up damaging a stone wall separating two fields. In most cases, Terraca and Obrisa fixed these problems (repairing damaged stone walls, compensating for unaccounted tree removal, etc.), but the incidents created a general climate of litigiousness and friction that was palpable during my fieldwork, when I could observe these clashes as they unfolded in the agricultural fields of the village. The disputes exemplified the sense that the developers mistreated the locals by disrespecting their possessions—as the people of Fatarella like to say, "They behave as if they were the landlords [*amos*]."

Third Movement: Accessing Land

When I arrived in Fatarella in 2010, the wind farms of Corbera and Vilalba were nearing completion. The initial construction work on the wind farm of Fatarella was underway, coinciding with the execution of the last round of expropriations for that facility. It was not rare to witness conflicts between company workers and peasants in the fields, who often made use of direct strategies, such as placing a tractor at the entry to prevent the machinery from accessing it. Yet no one talked about the wind farms in the village. That silence felt heavy.

My presence in the village broke that silence—and even then, just barely. Because most people knew that I was interested in wind energy development, my daily routines were punctuated with conversations about the wind farms. These comments, usually uninvited, tended to fit a repetitive formula.

It would go something like this: "Do you want to know what I think about the *molinets*/*ventiladors* [slightly pejorative terms for wind turbines]? That whoever has one is happy with it, and the rest... So I wish I had three."

On a few occasions, my presence broke the silence in a more dramatic way. In the winter of 2011, Mercè invited me to watch one the games of the women's Parcheesi championship. The women of Fatarella created this championship, played in couples, in the late 1990s to combat the isolation of winter nights. As Mercè told me during the game: "We needed a reason to go out and this was it. Men played cards, so we needed a game. It seems like a stupid game, but it is unbelievable how much it's given us. We experience it with passion."

Carla, one of the competitors, added, "But it has also been a way for us to break barriers, because we lead different lives and we barely talk to each other, so this has been a way for us to get in contact." For the first twenty minutes, the game seemed to be performing its function: a precariously employed tourist guide, a civil servant, a supermarket cashier, and a farmer, ages ranging from the mid-thirties to the early fifties, married and single, played competitive Parcheesi late into the night, sharing laughs and stories, talking about work, families, and the economic crisis in a friendly way.

But then Mònica, the only player I had not met before, asked me about the *molinets* and the whole atmosphere of conviviality was shattered. The couples argued in an increasingly antagonistic tone, the conversation devolving until their different positions were difficult to discern. After five minutes of escalating reproaches, the women decided to stop the argument and politely asked me to leave.

Beneath the village's code of silence, there was a poisonous conflict. Some villagers compared that silence to the postwar period, marked in Fatarella by the *Fets* of 1937 (as we saw in chapter 1, more than thirty villagers were executed with the alleged complicity of relatives and neighbors). The comparison is hyperbolic, no doubt, but unsurprising. As in the postwar period, intense household discussions contrasted with a steely silence in public, and the locals' relationship with this public silence was ambivalent. Everyone seemed to understand it as a necessary mechanism to protect public relationships from fracturing, but it also perpetuated an underlying rupture: the breach of *convivència* that the municipality had tried to preserve in the initial stages of wind farm development.

Everyone shared the feeling that the process had created losers and winners. More importantly, as villagers used to say, using a peculiar idiom, wind farm development had turned the village into *forest*: an unstructured social universe in which everyone looked out for herself and factionalism reigned supreme. This silence left space only for sudden outbursts of conflict—as in the Parcheesi championship—or for an uneasy ambiguity.

Formulaic comments such as "I wish I had three turbines" inhabit this ambiguous space. Taken at face value, they suggest both enthusiasm toward wind farms—the more the better—and a complaint about their scarcity—not every plot of land has a turbine. At a deeper level, the ambiguity is subtler, for the statement could express either a defense of selfishness or a critique of it. And yet, both possibilities pointed in the same direction: collective responses had failed, and selfishness was the outcome of wind farm development in the village.

As we have seen, some level of conflict had always surrounded wind farm development in Terra Alta and Fatarella, but 2007 marked a turning point. Until that moment, the conflict was between wind energy companies and locals. The conflict spilled over into Fatarella's social life as a consequence of two concurrent processes: the emergence of an internal political party conflict and the beginning of Obrisa's negotiations with the landowners affected by its three wind farms. Compared to the negotiations for Terraca's line, this new round involved not only a quantitative difference—close to three hundred affected landowners—but also a qualitative one, for coveted land leases coexisted with less desirable contracts for the rights of access of associated infrastructure. This uneven distribution created internal competition and conflict, further spurred by Obrisa's negotiation strategies and the village's internal political division. The precise relationship between these two conflicts is the subject of contrasting interpretations that divide the village to this day, yet there is little doubt that their interaction escalated conflict and division.

The Political Conflict

In 2007, the independent party that had controlled the village's municipal government since the end of the dictatorship lost the municipal election to Esquerra

Republicana de Catalunya (ERC), a left-leaning party that had been dormant in the village for decades. ERC did not get a majority despite winning the election, and its victory was controversial because the party had received the support of Fèlix, the outgoing independent mayor and historical leader of Unió de Pagesos (UP). The election, therefore, signaled the split of the independent party and the dissolution of the progressive hegemonic power bloc that had emerged in the village after the dictatorship.

Feeling bitter and betrayed, the independent party loyalists speculated that Fèlix supported ERC (which at that point was a coalition member of the Catalan government) because of the wind development project. They suggested that he had a personal interest in the project. As we may remember, the Agreement stipulated that if in four years the wind farm was not built, the municipality could unilaterally rescind the Agreement, but Fèlix, apparently with little consultation, decided to extend the Agreement before he left office. Sometime later, the independent party discovered a juridical report advising against the Agreement; Fèlix had kept it secret. Allegations that Fèlix acted out of personal interest were never proved, yet it is clear that bureaucratic opacity and the Second Miracle's characteristic proliferation of corrupt practices provided a fertile ground for suspicion (see chapter 5).

Further, the controversy over land leases increased the polarized political climate in the village. This tension was particularly noticeable at the institutional level, with the three parties represented in the municipality—independents, ERC, and CiU—constantly presenting motions around the wind energy issue. In November 2007 alone, the three groups presented as many as four motions: demanding a new wind referendum, requesting information on the effect of electromagnetic fields of high-tension lines on the human body, opposing wind farm overcrowding, and demanding an immediate halt to the three wind farm projects. Because of the complex composition of the town council, all these motions were passed. Behind all this noise we find, in fact, three different positions: the independents wanted to paralyze the wind farm of Fatarella and push for a general moratorium on wind energy in the region; ERC accepted the projected wind farm in Fatarella, but wanted to show that they were against "wind farm overcrowding" and proposed legislation preventing any future additional wind farm in the village; and CiU criticized the whole process, but in fact wanted more wind farms.

Given the situation, the celebration of a new referendum seemed like a reasonable compromise. Obrisa was, however, strongly opposed to that possibility, since a negative vote would force the municipality to freeze the wind farm project (technically within its power, for Obrisa still needed the municipal construction permit). Alícia, the new mayor, sought the advice of the juridical services of the province. The response was unambiguous: denying the permit would mean unilaterally breaking the Agreement, and this would have dire consequences. The municipality would have to return the advance payments (at that point, three hundred fifty-nine thousand euros) and face the possibility of a lawsuit with "incalculable consequences." The referendum could not be held.

Although the juridical report made clear that the municipality had no mechanism to delay or to block the wind farm, the political polarization intensified. An association called Associació per a la Defensa de la Terra i el Vent de Tots [Association for the Defense of our Land and Wind], popularly known as Ventdetots, was created in early 2007, before the election, by a group of landowners who sympathized with the independent party. The group campaigned tirelessly against the wind farm. In 2008, Ventdetots organized a demonstration through the streets of the village with dozens of attendees and filed a lawsuit against Obrisa, finally dismissed in 2009, for entering properties without the consent of the owners. Many villagers resented these actions, believing that the independent party was using Ventdetots to erode the municipal government in order to regain control. Wind energy developers were able to tap into this internal division to play residents' interest against each other.

Land Negotiations: Favors and Envy

Although much of the discussion around wind energy in Fatarella played out as a debate between different political parties, the conflict ran much deeper. Wind farms fueled old enmities, intensified factionalism, and antagonized neighbors. Conflicts emerged in the fields, where neighbors would argue over the precise limits of their property, even as they joined forces to blockade the wind company prospectors. They also emerged inside the household, with parents and children disagreeing about accepting the developers' deals. Roger, a farmer in his late thirties, recalled:

> One day my father tells me that they called him to say that we had a turbine in the fields we have in Corbera. "I don't want it, I just want to farm the land, without that contraption," I said. The conflict started right away, all because of these people [the wind energy company]. To stop the argument, we decided to have a vote, with my mother and my sister, because [things were] going bad, [we had] big arguments while having dinner, and we didn't let my wife and my brother-in-law get involved, because it would have been worse. We voted and I lost, and since then we haven't talked about it again… True, they give you 3,400 Euros [a year], but it is something I didn't want, it is a calamity. I mean, I see my father's point of view, it's money and [if we didn't take it] they would have put it in the neighbor's, but sometimes I tell him: one day, I may be eighty, but one day that thing will come down.

Obrisa's strategy for accessing the land contributed to these clashes. Indeed, early in 2007, while the procedures for the expropriation associated with Terraca's line were still ongoing, Eva and Eugènia started to negotiate the land leases for the turbines. This concurrence meant that Obrisa pitted two kinds of landowners against each other: those who received a land lease and those who were to host the associated infrastructure. The latter had to accept the modest deal offered by the company or face expropriation. On the wind farm within Fatarella's municipal territory, thirty-two of the one hundred ten affected landowners rejected the land contracts and ended up facing expropriation once the wind farms received the designation of public interest. The former knew that resistance in the form of rejecting the lease was futile. As the company would frequently tell them, the engineers could simply move the turbine to the neighboring field; there would always be someone willing to host it.

Nonetheless, resistance to signing land leases was rare, for these leases involved a sizable economic compensation. To make the sale more attractive, Obrisa offered advanced cash payments corresponding to half of the first annual lease. More importantly, those favored with a turbine saw some co-villagers' resistance to the project as a threat to their revenue. It gave rise to all sorts of dirty maneuvering, Eva remembered: "One of the families who had a turbine also ran a bar, and you know, in a bar you hear about everyone. So they would call us and tell us about those who complained more and with whom we had to be more careful, or whether they [the "anti-wind" group] were planning new activities, and all that."

Turbines were a limited good. In Fatarella, there are close to four hundred landowning domestic units (including around eighty landowners who do not live in the village). Although the wind farms affected most of these domestic units, there were only sixty-three turbines in the three wind farms, fifty-eight of which corresponded to fifty landowners of Fatarella (eight of the landowners had two turbines). Those who had no land affected by the wind farm voiced their grievances in a recurrent expression: "I get no money, but I see the *molins* anyway."

The limited number of turbines was, in fact, the basis of Obrisa's fear of losing an eventual referendum, as Eva explains:

> If you do a referendum when nobody knows about the specifics of the wind farm, who is affected and who isn't, then that's a clean referendum. But if you do it afterward, people are already starting to think, "Look, my neighbor is getting 3,000 Euros a year for doing nothing, and I see the turbine anyway but don't get anything?" You cannot do it like that. When you are distributing the raffle and all the numbers are in the hat, then everyone buys, but once the awards are distributed you cannot do a referendum, because you are going to lose it.

Here, the raffle is more than a simple analogy; it discloses the logic through which the wind developer interprets the behavior of the local population. Favors and envy are tightly integrated. Obrisa framed the turbines as a scarce prize or a favor that the local landowners should not and could not reject, as the following quote from Eugènia shows: "Those who have a turbine should be grateful, because what that person is going to make from that field she would never make otherwise. Do you think they make 4,000 Euros from that? I told them: take this money and you'll see how much you'll have for yourself once you retire. You will be able to go live in the Bahamas!" For Obrisa, resistance to the wind farm could only be understood as the result of envy or political manipulation, both of which were nurtured by the poverty and ignorance of the local population. For wind developers, local opposition finds its root cause in misery, a misery that is economic but has moral and political effects: because the locals are poor they are envious of their neighbors' wealth and this envy makes them easy prey for (selfish) platforms and opposition groups.

The reality is, of course, far more complex. Yet during my research I found plenty of evidence to suggest that the local population shared some of this interpretive framework, as the repeated comment "I wish I had three" suggests. Locals, for instance, internalized the idea that the turbines were favors. In early 2012, after the first wind farms had begun operating, many would refer to the annual payments they had just received for the first time with expressions such as "I received the visit of three wise men," a reference that could translate as receiving a present from Santa.

There is also ample evidence that in Fatarella the distribution of turbines led to resentment and factionalism. Take, for instance, how Gonçal, CiU's mayoral candidate, argued in a town hall meeting against the proposal to prevent the possible installation of more wind farms: "If we only wanted one wind farm, it should be located in public lands of the village. This is how we could get a benefit for the whole village, for we would be able to pay, through the municipality, the electricity bill of all the residents. Also, who are we to say that some people can earn 3,400 Euros and others can't? How is it possible that some council members who have relatives with turbines or who receive other high leases can suggest that we prevent other people from receiving the [same] compensations?"

The line separating personal and group loyalties from party allegiances is always blurry in small Southern Catalan villages. The importance of family and friendship networks, the role of the municipality in providing jobs, and certain forms of dependence (for instance, between business owners and employees) give political parties a fairly stable and cohesive structure, creating a permanent risk of them acting as corporate factions. Gonçal's comment shows that the generalized climate of suspicion pervading the actions of the municipality, the political breakdown, and the uneven distribution of turbines created the perfect mix to push that unstable equilibrium off-balance. Envy, group loyalties, and entrenched personal enmities started to dominate the discussion around the wind farms, largely confirming the vision of the wind development company. The passions for divergence became stronger than the passions for convergence that underpinned the cherished and fragile *convivència*.

Beyond Envy

All this notwithstanding, the developers' interpretative framework, structured around envy, favors, and misery, has serious limitations. Perhaps most evidently, this framework conceals the fact that in many cases resistance to the wind farm project originated as a direct reaction to the actions of the developer, most notably their widespread threat of expropriation.

Second, and more importantly, this framework contains a crucial contradiction: the economic misery of the locals, presented by the developers as the root cause of opposition, was not accidental, but central to the planning of the wind farm. It made access to land cheaper and easier, given that it was an economic opportunity that locals could not reject. Eva learned this while trying to negotiate the land leases for three wind farm projects that Obrisa had in the wealthy northern Catalan county of Empordà. Eva suggests it is this wealth that prevented the construction of those three wind farms:

> Have you been to a village in Terra Alta and then to one in Empordà? In Empordà they have money and they couldn't care less if you pay them a bit more or less. But in Terra Ata they need the money. So in Empordà they can afford to despise you—in Terra Alta not so much. I mean, have you been to Agullana [a village in Empordà]? It is gorgeous, it has amazing historical houses. So the guy who lives there does not want a turbine and then have to have arguments with his neighbors. Why should he want it? "I already have my beautiful house, my garden, why should I [have to deal] with all this mess of the turbines?"

The capacity of well-to-do regions successfully to oppose wind farm projects ends up reinforcing the developers' preference for peripheral, impoverished regions. This generates a tautological dynamic by which wind farms become markers of peripherality, economic marginality, and political passivity, often perceived by the wider public as what Andrew Blowers and Pieter Leroy call "locally unwanted land uses," which "tend typically to be located in already backward areas and, therefore, their location reproduces and reinforces processes of peripheralization."[24] As Eva says, the inhabitants of Terra Alta "need the money" and cannot reject the economic offer. And this is true even if, as they say in Fatarella, "it is not enough money to fix anything," an expression that elliptically compares the land leases from the wind farms with the jobs from the construction of the nuclear plant.

Fatarella's landowners were keenly aware that Obrisa benefited from and enforced their peripheral condition. Let's take the case of Bernat, who had been deeply involved in the negotiation of the Agreement. In 2007, his uncle—old, single, childless, and, according to his nephew, "an awkward man"—was offered a land lease for a turbine, but Obrisa insisted that he needed to register his title in order to sign the contract. After a series of meetings, the family decided to transfer the property to Bernat and register the land title. A couple of months after the initial contact, Bernat went title in hand to see Eva and Eugènia in the municipality, where all the contracts were signed. There, Eva told Bernat that Obrisa had had to move the turbine for technical reasons. Bernat did not believe it: "I think I know what happened, but I won't tell you," he said to me.

"Did it go to a neighboring landowner?" I asked.

"Kind of neighboring. I suspect that a favor had to be paid."

Instead of the turbine, Eva told him that now his land was affected by a ditch, and the company was offering to pay a thousand euros for right of access to a thousand square meters of the uncle's land. He then told them:

> One euro per square meter? Do you think this is the right price [*el preu que toca*]? Okay, look, since a thousand euros won't fix anything for me [*no m'arreglaran res*], I won't get rich, and after all I don't want money, you know what, let's not put money on the table, let's not get angry, we'll make it simple: we'll barter [my thousand meters] for a thousand meters in Barcelona [where Obrisa is headquartered]. They got livid. They said that this could not be done, they are not worth the same, they said.... I came back: who says that? Why not? Who sets the price? Is it you? Because if that is the case then there is nothing to talk about, but I don't understand why you have to set the price.

Bernat's response denounced the false character of the negotiation, for as it is said in Fatarella, most often to refer to agricultural prices and the small producer's inability to set them, "It's only business if two people do it; otherwise it's called something else." His reaction traces a link to a historical experience of dependence while exposing the political character of land value measurements: value does not appear as a neutral indicator of market dynamics, but as the outcome of peripheralization and imbalances of power. Bernat ended up selling the piece of land back to his uncle, who died a year later, bequeathing all his property to the church.

In the third place, the developers' comprehension of the local opposition to wind development neglects the fact that the action of the political parties was not only oriented to obtain personal and group advantages and that most villagers criticized wind farm development on communalist grounds.

The villagers' communalist vision was most often articulated as a demand for free electricity, such as "energy should be public," "it doesn't make sense that we make more electricity than we can consume and we still have to pay for it," and "we should all get the electricity from the wind farm, no electricity would be lost in transportation and we'd be a truly ecological village." The first comment was frequently repeated, signaling a clear opposition to the regulation of a public service through market mechanisms and private interests.

Taken altogether, these expressions expose and interrupt the asymmetries on which wind energy production rested: the internal division between lucky and unlucky, the uneven power position of locals and developer, and the division between poor, peripheral regions producing energy and core areas receiving it. These pervasive comments and criticisms show that despite the undeniable intensification of factionalism and personal clashes, Fatarella's inhabitants never lost sight of the structural contradictions of wind farm development in Spain: uneven ecological distribution, separation of production from consumption, weight of corporate interests in the design of the electricity system, and so forth. Be it through the ideal vision of publicly owned wind farms serving local interests or through the widespread practice of installing small turbines and rooftop solar panels, the people of Fatarella offered an alternative vision of wind energy production.

Nonetheless, motions, enmities, demonstrations, lawsuits, blockades, envies, arguments, and complaints did little to prevent Obrisa from accessing the land it needed. In 2009, the wind farm of Fatarella was designated to be of public interest (the other two were designated in 2008), allowing for expropriations to be issued in order to overcome the final obstacle to securing access to land. A few months later, the wind farm received its last governmental authorization. At that point, before a single turbine was installed, Obrisa left the scene.

Fourth Movement: Congealing Value

In 2009, Eudald sold Swellwind, the subsidiary that owned the three wind farms, for 28 million euros to Energias de Portugal (EdP), a formerly state-owned Portuguese utility and one of the members of the oligopoly. The news, announced only once Obrisa had obtained the last municipal permit required for construction, baffled Southern Catalans. On the one hand, the contrast between the money Obrisa had received and the economic compensation offered to the landowners and municipalities was seen as abusive. On the other hand, the intangibility of what was being sold triggered complaints of speculation and financial corruption. As the locals like to say, Obrisa "just sold paper": permits, the Agreement, and land leases. While it was a less important concern, the Portuguese origin of EdP only added to the locals' indignation and gave ammunition to the platforms, which posted road signs in Portuguese.

"This is Portugal now," some said, a complaint that would become "this is China now" in 2012, when the Chinese Three Gorges Corporation became the majority stockholder in EdP's wind branch.

Eudald took all the necessary steps to access wind. All that was left was to harness it, making electricity and profits. This task would be left to EdP. The 28 million that Eudald received may be considered an advance on these profits or, if you like, his cut of the business. Right after the sale, EdP started building the three wind farms, which went into operation between 2010 and 2011. EdP owns three other wind farms nearby, creating what looks like a semi-continuous line of turbines extending over twenty kilometers. Regardless of its nationality, everyone seems to agree that EdP "has become the landlord of the territory."

Even if one accepts that Eudald did not plan to sell the three wind farms, as he insists, he had taken the precaution of completing the steps needed to make that sale possible, both by including that possibility in the Agreement and by creating Swellwind. The sale is especially paradoxical if we also remember that the main reasons that led the municipality of Fatarella—and, by extension, Corbera and Vilalba—to reach a deal with Obrisa was its local character and relatively small size. Indeed, the three municipalities lament the sale and believe that it was easier to sit down and negotiate with Eudald than it is with EdP.

Eudald presents his sale as the result of a financial coup against his company. He had accessed the grid, the permits, and the land, but then found unexpected problems accessing capital. With the beginning of the 2008 crisis, credit dried up, and banks were pressuring Obrisa, presumably heavily indebted: "They were putting their hands in my treasury, and they make you do whatever they want." In other words, the banks forced Eudald to get cash by selling some of his assets: the three wind farms in Fatarella. Although perhaps not in the way that he had anticipated, his wind energy adventure did ensure the sustainability of Eudald's family construction business.

He argues that Obrisa was the victim of credit restrictions and the oligopolistic utilities' drive to expand and control the wind energy sector, for which the financial situation provided a perfect context (see chapter 5). Indeed, Eudald had no problem finding a buyer. Large utilities' aggressive strategy to increase their control of the wind sector created a great deal of interest. Whereas Spanish wind energy development was at its peak between 2004 and 2007, in Catalonia the regional government's lukewarm support for the sector and the intense popular opposition to it delayed the process. Thus, until 2008, Catalan wind energy was almost entirely dominated by medium-sized, independent developers. Around 2008, however, multiple wind farms received governmental authorization, converting Catalonia into one of the regions spearheading Spanish wind energy development. Gas Natural and EdP, members of the oligopoly, used their size and financial power to take control of Catalan wind, with the purchase of Swellwind as a centerpiece of the strategy. Eudald explains: "I went to Madrid to talk with one of the big [electricity companies]. If the bank was offering X for the wind farms, this company doubled the offer. It was an electricity company [EdP] that had to grow in Catalonia, because they had never had a presence here and they needed that presence. For them that was all. For us, building wind farms was a matter of philosophy. We thought that as a family company moving into the next generation we could improve society: [create] a society without foreign dependence, producing clean and autochthonous energy."

The oligopoly's expansive drive in the early stages of the economic crisis may be described as what Frédéric Lordon calls an "entrepreneurial conatus," in which profits become secondary to the goal of expansion, a power struggle in which swallowing is the alternative to being swallowed.[25] Yet Eudald's profits were not only based on EdP's overpayment, but also, more crucially, on the small amount of capital that he had had to devote to manufacture the intangible commodity that he sold.

He admits as much: "I mean, I don't think drug trafficking is as profitable. The business is lucrative, certainly. But it was lucrative because we did it ourselves, because I put in my own labor plus the labor of engineers who were already part of my company, who were already part of my personnel costs. *You go see the peasant every two weeks and that's it.* The cost of the administrative permit is close to zero, and if it is zero then the yield on investment is infinite." Eudald made something for nothing.

After the Silence

In 2013 the village seemed to recover a semblance of normality. The crews of wind farm workers had long since departed, and the silence around wind energy was less marked, even if the scars of the internal conflict were still apparent. In September, the village was celebrating the *Festa de la Misericòrdia*, parading the Virgin of Misericòrdia's image through the streets and singing religious tunes in varying degrees of sobriety. Like all celebrations, it included a Saturday night dance in the *casal*, which was not to start until close to midnight and was preceded in every household by a special, lengthy dinner with friends and relatives.

One of my neighbors, Pere, invited me to join the dinner that he and his wife, Meritxell, had organized. It was attended by half a dozen of Meritxell's friends and relatives: her sister, her cousin, and two friends, Berta and Mari, who came with her husband. This circumstance is very common in Fatarella, and the locals use a cryptic idiom—"the friends are hers"—to express the fact that most groups of friends, called *colles*, are structured through female networks. I had learned that in Terra Alta you must never bring a bottle of wine to a house dinner, for someone will bring wine that she herself made, but this was especially so since Pere and Merixell process their own wine and have a modest company selling it in bulk to bars and restaurants. Pere also likes to make all kinds of spirits and dessert wines—*ratafia, mistela, vi de missa, vermut*—and he put these on the table after we ate.

Over the drinks, my presence prompted Berta to start a conversation about the wind farms. Although she visits Fatarella frequently, Berta had moved to Barcelona decades before and was the only person at that dinner that I had not previously met.

She rushed to reiterate a point I'd heard many times: "Do you want to know what we think about all this? That whoever has a turbine and makes money is happy, and the rest aren't. As simple as that." This time, however, the sentiment was met with annoyance. The other guests disliked Berta's depiction of the villagers as selfish and resentful, though none had been particularly opposed to wind farm development.

The conversation that ensued was thoughtful, long, and, at times, tense. One specific exchange powerfully condensed the tensions and contradictions that wind farm development brought to the village. Almost everyone at the table agreed that the process of wind farm development had not been carried out in the proper way: "We should have created order [*posat ordre*], and instead of that we've been fighting against each other." This idea returns us once again to the dialectic between center and periphery, the capacity of the center to disorganize the periphery, to export disorder into it. If order refers to the local capacity to govern the development of wind energy—also the main demand of the platforms—and adjust it to local reproductive needs, development had created disorder and transformed the village into a forest, unleashing an insidious infighting.

Mari, who is a nurse, and her husband Enric, an electrician, interjected that the cause of that infighting was that everyone "swallowed the bait and thought that she would become rich, and that's not what's happened."

They traced a connection, obvious to everyone at the table, to Southern Catalan farmers' historically subordinate position in all manner of dealings, particularly where they were unable to negotiate prices and therefore to control their means of reproduction: "They did the tactic of divide and conquer, and it worked, because they told the people in the county that you had to accept whatever you were offered, and this is why we are poor. It is because we are poor of spirit that we are poor in our pockets, and also the other way around."

Pere agreed with Mari's argument, but disagreed with her tone: "Yes, people were misled, and no one will get rich, that's all true, but if you get 3,000 Euros a year, that's quite something. With the almonds and the olives we can't make a living here. Three thousand Euros would have gone a long way for me. You, Mari, know that you get a check by the end of the month, but that is not the case for me."

Mari said that, with recent massive cuts in the public sector, it was not so clear anymore that she would get a check.

"Well, that's the difference," said Pere. "For you it's new, but for me it's always been like this. Also, for me the *molins* are not a curse [*la pesta*], the curse is [the nuclear plants in] Ascó. And for me the real ecological disaster is not seeing Ascó or the wind farms or any of this, but seeing untended fields going to waste [*camps sense treballar que se fan erms*]."

Joana, Meritxell's sister and the eldest in the group, ended the verbal exchange: "Let's be clear, if the people have put turbines in their fields it is because there is no generational changeover. Households are disappearing, the youth leaves, so since there is no expectation for the future, just give me the money."

There is no unified logic to the disagreement between Mari, Pere, and Joana, but there is a shared experience of resistance and precariousness that is fundamentally expressed in relation to land and farming. Everyone agrees that the main problem—the "real ecological disaster"—is the stagnant situation of local agriculture and the resulting loss of population. Everyone agrees that there is an intimate, two-way relationship between being economically poor and being "poor of spirit," an idiom closely connected with the idea of dignity, the capacity to assert one's worth, to govern your own destiny, to be master of yourself. The drama of wind farm development in Fatarella is that it has not solved any of these problems. And while it is true that neither has it created them, it has made them evident, eroding the local sense of self-worth, disorganizing the social fabric, and leaving many wondering whether the company was perhaps right—perhaps their land, and with it themselves, were worth less than the annual leases.

Notes

1. Ribot and Peluso (2003: 154).
2. Wolf (2001: 384–385).
3. For the realization of this section I benefited immensely from the nearly unlimited access the municipal government of Fatarella gave me to its archive. Oleguer Presas' (Presas 2014) and Sergi Saladié's (Saladié 2013) unpublished reports have also been greatly helpful.
4. Caffentzis (2013: 153).
5. Hornborg (2001).
6. Barthes (2015).
7. See Caffentzis (2013: 177–179); on thermodynamics, see Smil (2008) and Martínez Alier (1987).
8. Hornborg (2011); see also Coronil (1997).
9. Quoted in Caffentzis (2013: 153); Marx (1976: Parts Three and Four).
10. Fatarella's location is excellent: besides guaranteeing upward of three thousand hours of operation per year, Fatarella's wind is remarkably nongusty, thus causing few mechanical problems.
11. For a short summary of Obrisa's trajectory, see Canosa (2014).
12. See, for instance, "El corralito catalán," *El País*, March 6, 2005.
13. My description of the terms of the Agreement is based on a revised version from 2005.
14. Taxes fall outside of the scope of this chapter, for they only come into play with the start of operation of the wind farm. I should however note that taxes provide sizable municipal revenue and are a recurrent source of contention with the developer. In Spain, wind farms have to pay three kinds of taxes: construction tax (called ICIO), business activity tax (IAE), and real property tax (IBICE). IAE is calculated as a percentage of the sale of electricity, reporting around fifty thousand euros annually to Fatarella. ICIO is a one-off tax calculated as a percentage (between 1 and 3%, depending on the municipality) of construction costs.

IBICE is calculated as a percentage of the value of the fixed assets (most notably the turbines); since 2012, the municipality has been litigating with the developer to determine that value.

15. Presas (2014: 48).

16. Presas (2014: 50–51). From 2006 to 2012, Eudald also was the president of EolicCat, the Catalan association of wind farm developers.

17. Lleonart et al. (2006: 29).

18. The two uncompleted wind farms are located in Orta and Riba-roja. Orta's project came to a halt as a consequence of local opposition and it is unlikely that it will ever be built; in Riba-roja both the mayor and the developer (also Obrisa) hope that the project will be completed in the near future.

19. Plataforma Terra Alta (2007), last edited December 15, 2007, last accessed October 3, 2017, www.plataformaterraalta.com/manifest%20mani%20tortosa.htm.

20. Sempere (2008: 143).

21. The quote is extracted from an interview Andrés gave to the very popular television program *Salvados*, included in the episode "Lobby feroz," which aired on March 17, 2013.

22. Lleonart et al. (2006).

23. All the cases I know of have one thing in common: the dwellings belonged to Northern European families who had bought them as second residences during the real estate boom of the miracle.

24. Blowers and Leroy (1994: 198).

25. Lordon (2002).

7 Waste and Dignity

My struggle is a struggle for survival, and for all survival to have a place, without that place depending on someone wanting to give it to you. Goats should have a place to graze and humans a place to survive.... But that place should not be measured, unless it is oneself who has the meter, the inner one, life's meter.

—José Domínguez Muñoz, *El Cabrero, el Canto de la Sierra*

I HEARD THE knocker landing heavily against the robust wooden door. From the doorstep, a tall, corpulent man identified himself through his house name: "It's Daniel de Figuera." I was waiting for him. The day before, Daniel had told me he would drop by "after lunch"—*sobremesa*, that indefinite moment of the day when life slows down in Spanish villages, the streets oddly silent as Spaniards pause for lazy conversations, siestas, and card games in the bars. I had been meaning to talk to Daniel for quite some time. Villagers had responded to my difficult questions about local history time and time again with, "I am not sure. You should ask Daniel." In Fatarella, Daniel was unanimously considered a wise man. As I would soon learn, that is why the wind developers had sought his help.

It was a sunny, autumnal day, and Daniel suggested that we talk on the terrace. He grabbed one chair with his big hands and knuckled the seat: "Nice chair: made out of *boga* [cattail], by hand, probably in one of the river villages. They are hard to find nowadays." I wanted to tell him a bit about my research, but he seemed uninterested. I had already filled him in over the phone, and besides, everyone in the village already knew who I was: the anthropologist who wanted to know about the wind farms and the nuclear plants. He did not show much interest in my questions either; he preferred to tell me, at his own pace, who he was.

Daniel was born—as was the Civil War—in 1936. He spent the three years of armed conflict living with his mother and relatives in Barcelona. On their return to the village, Daniel and his mother found their house in ruins. His father was held in a prison camp in Lleida—about a hundred kilometers to the north—for several more years: "It was the mayor's fault, it took him a long time to send the *aval* [support letter]. And the priest didn't help either. I don't know if you're Catholic, but, because of that, I can't be."[1]

Daniel started farming with his father when he was twelve, "But the fields were full of bombs and corpses; it was impossible to farm. So my family sent me to work the land of an aunt: it was sheer exploitation, but they fed me, so I don't complain." Meanwhile, his father worked as an agricultural servant in a strong *casa* of a neighboring village and visited the family every two weeks. "But little by little we cleaned the fields. I mean, the rubble was part of the harvest, any money that the young people had in our pockets came from that: today a piece of metal, tomorrow a bullet." They even had love songs about gleaning the scrap: "If you send me a letter I'll tell you my whereabouts // I come from Fatarella, recovering mortars."

After a few years spent clearing the fields of debris, Daniel and his father began working their own land. They had seven noncontiguous plots (totaling around ten hectares) with olive trees, vines, and almond trees. Olive oil and wine were sold through the local cooperative, keeping a part for household consumption; almonds were sold to local traders for export. They kept a pig, rabbits, and chickens, as well as a vegetable garden and a patch of forest for firewood.

Daniel never married and never worked elsewhere: "I didn't go to the nuclear plant. I was doing well with my land, plus I had no children and [I had] my father's pension."

Currently, he works only a few hours every day in just two of his fields: "I go to my land in the morning, for a few hours. I don't need much. In the afternoon I go to the café and do some other things. I wouldn't even need to go to the fields at all, because I have no one to maintain, but I want to maintain the land, to preserve it [*mantenir-la, conservar-la*]. This is a very difficult land, and if you don't maintain it, it very quickly becomes forest [*se fa garriga de seguida*]."

Only after he had told me of his life was Daniel ready to talk about the wind farms. He started, as is almost always the case in Southern Catalonia, by putting wind energy development in a longer historical context of energy production: "With the nuclear power plants, we were cheated with the 'chocolate factory' and all that nonsense; we didn't know what was going on. And with the wind farms it's been the same. We couldn't negotiate. The right conditions for negotiation were not in place."

His own relationship with wind developers was an unsettling experience to which he would return every time we talked during the following years. Given his deep knowledge of the area, the medium-sized, Catalan firm Obrisa hired him in 2006 to help them build the electric line that would connect the wind farms to the long-distance grid (see chapter 6). For months, Daniel accompanied Eva and Eugènia, helping them identify the owners of the parcels of land affected by the line. He was a sort of double agent, on several occasions claiming ignorance to the developers while later warning fellow villagers: "Hey, the wind people want to build the line over your field. Go quick and register your title."

In 2007, when Obrisa was starting to negotiate the land leases and conflict was mounting in the village, a comment from Eugènia spurred Daniel to quit working with the developer:

> When we were almost done, the lawyer [in fact: the engineer] said to me: "And you? I see that you don't have any plot of land with a turbine." She was suggesting that if I had told her maybe we could have found an arrangement, but instead of going that way, I responded, "No, I don't, and I am happy not to have any turbine in my property." You know? I said that because if I don't have turbines I am still the *amo* of my land ... once they put a stick they are in control, even if the land is still yours. But then she said to me: "But how come? These lands are worthless and now with the wind farm you could get some money." And this was too much! "Do I look hungry? I've lived out of this land all my life. It's you who thinks they are worthless." And she came back, "I don't understand how you all keep resisting with these miserable lands that you have." "Since the end of the war," I told her, "we've been living from this land, feeding all the mouths around the table. If you know how to farm these lands, they can give you a lot. But you need to know [how]." This is what I told her.... But now I would tell her that I want her to put one turbine in every one of my plots of land.... Look, they do the wind farm, regardless of us wanting it or not. Those of us who don't have a turbine are screwed, because we see them anyway.

Eugènia's words insinuated that Daniel's land, and by extension all his life, was worthless. All that ensued would confirm to him that the wind developer considered local land and its inhabitants worthless: the expropriations, the impossibility to negotiate the terms of the land agreements, the careless trespassing, the threats to move turbines to neighboring fields, the insidious suspicion of nepotism, the neglectful destruction of stone walls, the pervasive suggestion that locals were only moved by envy. Thus, to locals, wind energy development looked like old wine in new bottles, the extension of a centralized, extractive, concentrated model of electricity production, just one more episode in a long history that has reproduced the land and its dwellers as what Vinay Gidwani calls *waste*—obstacles to the march of progress and the realization of value.[2] Southern Catalans have historically reacted to this ongoing threat of peripheralization and marginalization, to the intimation that they are redundant, with an indignation that supports a fight for dignity.

The inhabitants of Terra Alta tirelessly complain about their difficulty in making a living, the impossibly low agricultural prices, the danger of nuclear radiation, the scarcity of alternative sources of income, and the disrespect that energy companies show for their land and their lives. They habitually punctuate these complaints with a final remark: *aquí sobrem*. It means, roughly, "we are redundant" or "we are a burden." The locals have two basic options: to surrender to this logic, accepting that they are waste, or to fight against it, asserting the value of their land and preserving the reproductive relations established through it, a resistance that takes the form of a demand for dignity. This concluding chapter examines this dialectic between waste and dignity, between resistance and surrender, in a historical moment where the contest seems especially imbalanced. With its mix of bitterness and irony, Daniel's final comment suggests that the energies to resist are running low, that soon it may all be over.

The inhabitants of Fatarella have not been able to deal with wind energy in a way that can be reconciled with their search for dignity. Indeed, wind energy development has had a profound and paradoxical effect. Few locals would deny that the nuclear plants were far more disruptive than the wind farms or that they represented a bigger threat to the reproduction of local livelihoods. Yet the nuclear plants also provided the political spark and the economic resources to fight against them. The wind farms are far more benign, yet also more insidious; they constitute a difficult and divisive political target and provide a welcome economic incentive, but not a life-changing one. And while the nuclear plants were always a bit far away, the wind farms have impacted the intimate texture of the local fabric, penetrating the fields and creating competition between households.

More decisively, wind energy development has taken place in a moment of heightened despair: a corrosive economic crisis, the lack of a political reference once provided by Unió de Pagesos (UP), the duress of the accumulated effects of five decades of lost battles, exodus of the most able, and increased economic marginality. Many believe that wind farms are the final nail in the coffin, as one villager once expressed with a graphic metaphor: "This used to be a cemetery; now they've put up the crosses." And yet, the search for dignity persists.

Waste

Vinay Gidwani's inquiry into waste extends for more than two decades. In his initial contributions, Gidwani analyzed the legal and normative category "wasteland" as used by the British colonial administration in India, showing that, in identifying certain lands as unproductive (i.e. yielding no revenue), this morally and politically loaded category justified dispossession and displacement. Waste(land) thus appeared as a specific instantiation of the Lockean notion of *terra nullius*: an unproductive land that capital can and should redeem through development, filling it with new subjectivities and relations of production.[3]

Gidwani demonstrated how the concept quickly extended from land to (inappropriate) human activities: "The term 'waste' beg[an] to describe not only types of lands but also a range of human activities that ha[d] brought these lands to their current state."[4] The legal and moral apparatus constructed around waste effectively rendered those human activities inappropriate and allowed for the unfolding of primitive accumulation. The practice of naming both humans and land as waste was rooted in their connection: unyoked, land would be ready for capitalist exploitation, and people, once displaced and dispossessed, would cease to be a burden and could become free laborers, also ready for exploitation.

More recently, Gidwani has extended his inquiry, taking the notion of waste as a broader analytical tool for understanding capitalism as an economic, ecological, *and* moral order. Waste—as concept, matter, experience and metaphor—embraces a truly large field: the pointless, the ineffectual, the disordered, the irrational, the worthless, the excessive, the stagnant, and the improper. Gidwani's approach mirrors Mary Douglas' germane characterization of dirt or pollution as "matter out of place," that which poses a "threat to good order."[5] The cultural order to which waste is a threat is the order of capitalist development and its law of value: "'Waste' is the recurring other of 'value' ... the antithesis of capitalist 'value', repeated with difference as part of capital's spatial histories of surplus accumulation. I also suggest that by tracing the dialectic of value and waste, or the 'positive' that acquires its valence against the background of the 'abortive' or the 'retrograde,' we gain insight into how capital always draws its economic vitality and moral sanction from programs to domesticate and eradicate waste ... For projects of value, waste is 'an enemy to be engaged and beaten.'"[6]

Critically, waste not only operates negatively but also positively: not only is it the opposite of value, it is also productive of value. This double valence is salient in the environmental dimension of waste, as Ferry and Limbert suggest: "The double meaning of the word waste as unusable and polluting dross and as those parts of nature that are not made use of underscores th[e] normative aspect of the concept of resources."[7] For capital, nature that is outside of capital is always waste, be it as untapped potential—the raw matter of extraction and production—or as refuse.[8] This dual logic of waste is central to capital's mystification: by deeming that which is outside of it an indeterminate disorder, value appears as the domain of *logos*. Capital emerges as self-reproducing value.

Thus, Gidwani concludes that the dialectic between waste and value structures the world into two kinds of people and two kinds of spaces. There are people whose labor is valued and rewarded, living lives seen as worth cultivating, and there are people who are of little consequence to global capital, living lives that are cheapened, used, and easily discarded: "Valuable lives, wasted lives, and mapped onto these, valuable spaces and spaces designated as wasteful. Colonizing and remaking wasted spaces as valuable spaces, excluding from political citizenship those whose labors are not counted. This is the continuing juggernaut of enclosure."[9]

The analysis of waste allows us to understand the permanent operation of primitive accumulation, not as a specific historical stage, but rather, using Werner Bonefeld's neat expression, as the "constitutive premise" of all capitalist relations.[10] When it comes to the relationship between nature and labor, this constitutive premise gives place to a distinct logic, grounded, in Robert Nichols' words, "in the appropriation and monopolization of the productive powers of the natural world in a manner that orders social pathologies related to dislocation, class stratification and exploitation, while simultaneously converting the planet into a homogeneous and universal means of production."[11]

Melissa Wright's ethnographic analysis of female Mexican factory workers offers a powerful example of the kinds of wasted lives produced by capital. "To view the Mexican woman as cheap and disposable because she has no skills and is not trainable is at once a construction of her as an embodiment of waste, even as her body provides the source of much value through her labor."[12] We may again appreciate the indeterminate multiplicity of waste. The female factory worker of the Global South becomes an untapped potential (waste) precisely because she is framed as surplus (waste), as cheap and expendable. Ultimately, she is made into a discard (waste). In other words, she can function as productive waste because she has been previously conceived of as disposable.

As we saw in chapter 6, the same is true of Fatarella's land: it is attractive not only because of the untapped potential that traverses it (wind), but also because it has been conceived of as disposable and is therefore cheap. That is, devaluation works as a precondition of capitalist expansion: because certain places and peoples are constructed as waste—residual, barren, marginal, disordered—capital can justify the need to intervene and make them valuable, that is to say, value-producing. Returning to Eugènia's comment to Daniel, wind developers treated Fatarella's land, at best, as a retrograde residuum, made valueless by local inhabitants' presence. They were a *burden* to the developer's access to value. The more the inhabitants of Fatarella attach themselves to the land, the more they keep resisting, in Eugènia's words, the more wasteful—ineffectual, irrational, stagnant—they seem and the more redundant they feel.

Gidwani concludes his argument by pointing to the limits of the law of value. Waste is not only the discarded antithesis of value, it is also the "placeholder for a vital logic that constantly threatens to evade and exceed capital's grasp."[13] The inhabitants of Fatarella constantly articulate that vital logic by expressing feelings of indignation and asserting claims of dignity. These claims are structured around land, asserting as valuable the link between place and people that makes them both appear as waste.

Land and Livelihood

Agricultural despair overwhelms the daily experience of Southern Catalans. These feelings are heightened in Fatarella, a village in some of the region's roughest, driest land. My field notes are punctuated by comments like "It is impossible to live off the land," often alongside tragic declarations: "We are surplus," "They should finish us off." Nonetheless, there exists the risk that these dramatic statements blind the visitor to the actual economic role that agriculture plays in daily

Table 7.1. Landholding households' relationship to agriculture in Fatarella.

	A	B	C	D	E	Totals
Resident	57	76	87	49	49	318
Nonresident	0	1	11	9	55	76

A) *Independent farmers*: Households fully devoted to agriculture; most also have additional sources of nonwage income such as public pensions;

B) *Semi-proletarianized farmers*: Agriculture is the main activity and source of income, complemented by additional sources of labor and income;

C) *Weekend farmers*: Agriculture is not the main income source; lands are cultivated as a "hobby";

D) *Sentimental farmers*: Households that do not cultivate their land, but contract labor to farm it;

E) *Withdrawn farmers*: Households that do not cultivate their land.

life, not only as a source of revenues, but also as the element around which livelihood strategies and reproductive cycles are structured.[14]

Some villagers helped me elaborate a survey to understand the relationship that the landowners of Fatarella have with their land. Through this survey, I identified the number of landowning households in Fatarella: three hundred eighteen of a total of three hundred sixty-one domestic units. An additional seventy-six landowners do not reside in the village, even if they own a house in it. Households can be divided into five categories according to their relationship with farming (see table 7.1). These are, I should note, emic categories, pervaded with moral attributes and judgments and expressing different local livelihood strategies.[15] They are flexible and porous, and the location of a household within one of these categories reflects a mixture of objective and subjective factors.

Sterile Agriculture

Independent farmers only work *al defora*, that is, they do not "go to the *jornal*" [literally, daily wage]. Historically, the independent middle farmer who did not need to *anar al jornal* represented a cultural ideal pervaded with notions of pride and dignity. Nowadays, the category evokes moral ambivalence: the cultural ideal butts up against a reality that is far from it. The category is seen as largely unviable, increasingly sterile, unable to sustain the reproduction of livelihoods.

Indeed, a look at the households included in this category makes it quite clear that agriculture, on its own, does not allow for making a living. At least thirty of these households are "sterile" (with no offspring or offspring that will most likely not reproduce): retired farmers with no offspring in the village, single men and, most common of all, households where an elderly retired couple lives with a single son (and in two cases daughter) past his reproductive age (fifty or older). Retirement pensions often play a crucial role in the economies of these households. Only thirteen households fully devoted to agriculture could be located in the "fertile" category—either nuclear or extended families with young children. In between, I found a series of fourteen *cases* that may go either way: childless couples of reproductive age living alone or with one spouse's parents. Should they have children, most of these fourteen households would probably move to the category of semi-proletarianized farmers.

The relationship between agriculture and "sterility" is complex. We could say that agriculture may well lead to nonreproduction, something that is especially obvious in the difficulties inheriting sons have in finding a partner.[16] However, the cause-effect relationship tends to operate in reverse: only those households that do not need to take care of young dependents can afford exclusive dedication to agriculture. I should also note that the agricultural operation of "sterile households" is not necessarily terminal: single sons living with their parents often cultivate very close relationships with the sons of their siblings with the more or less concealed hope that these nephews may take over in the future.

The data confirm the notion that it is impossible to live off of agriculture. Only a handful of households escape this rule, all of them middle *cases* (possessing a sizable amount of land, around twenty hectares) that have been able to expand their operations, either through quantitative or qualitative mechanisms. Roger—married, in his late thirties with one child—is an example of the former. He has expanded his agrarian activities by reaching agreements with other landowners to farm the most marginal land of the village—planted with hazelnuts—by deploying an intensive use of machinery, at least by local standards, and a variable amount of wage labor (generally two to five employees, always from the village). By commercializing the produce and, critically, accumulating European Union (EU) subsidies—especially important in the case of hazelnuts, as we saw in chapter 3—Roger makes a good living and generates a source of temporary employment for others. Although individual arrangements vary, as a general rule, the landowners do not receive anything in exchange, other than "a little sack of hazelnuts" and, crucially, maintaining their land as productive.

Pere and Meritxell, who have two young children, illustrate the qualitative strategy. By producing their own wine in bulk, they make a fragile but autonomous living, selling their wine to local restaurants and at weekend fairs. Antoni and Raquel, who have one young daughter, have been even more successful, turning to intensive techniques—such as handpicking olives—and organic production. "When we started with this everyone thought we were crazy, but it's working. Now everyone wants to do it, but I don't think they'll do it. You need to believe in it. It requires a lot of attention, smaller production, and targeted treatments. The best thing is that prices are stable."

When I first met them, Raquel and Antoni saw the possible installation of the nuclear waste storage facility in Ascó as a threat to their living: "How are we going to sell organic stuff made next to a nuclear cemetery?" The couple thinks very few farmers follow their strategy because very few believe in it. Yet one of the reasons local farmers do not believe in the organic strategy is that they see it under permanent threat by the possible installation of hazardous facilities. More importantly, the semi-proletarianized character of most local livelihoods makes it difficult to engage in this labor-intensive activity, which requires a continuity of care that is hardly compatible with temporary wage labor.

Semi-Proletarian Livelihoods

The second category is the one that includes the majority of farming households. It is characterized by self-exploitation and semi-proletarianization. Generally, the male adult is devoted to farming, often combining that activity with other jobs, which may be temporary (nuclear plant refueling, agrarian wage labor). Operation of a small business, such as a bar or a shop, where other family members also work, is also common; in Fatarella, there are as many as twenty-four retail stores (including bars), mostly run by households that combine this activity with agriculture. Other members of the household are wage laborers, most often outside of the village, and are generally construed as "helpers" in the agricultural tasks.

The amount of land and the quality of the complementary sources of income determine these households' financial situation. Some, in which the primary male is fully devoted to farming a sizable amount of land and the wife either holds a well-paid job (as a civil servant, for instance) or runs a modest but profitable small business, are fairly comfortable. Others may live on the edge of poverty, their semi-proletarian position an effect of their inability to find stable, well-paid jobs. Most semi-proletarian households may be found somewhere in between those two poles.

Let's take Bernat's case. Up until the mid-1990s, Bernat combined a part-time job as an electrician with temporary agrarian wage labor and farming his household's land with his father. When his father died, Bernat became the *amo* of his land and left his part-time job, but not the temporary agricultural jobs. He married Gemma, who worked as a supermarket cashier, and they had a daughter. In 2011, Gemma lost her job, which led Bernat to find a new source of income. Since 2012, Bernat has been working as an electrician for a subcontracted company in the refueling of the nuclear reactors, an operation that takes place every nine months and lasts for about six weeks.

During the refueling periods, Bernat works seventy-hour weeks and earns a considerable sum (around four thousand euros total). Of course, this job reduces his capacity to take care of his land, to the point that most years Bernat cannot harvest all of his produce by himself. "I can do two things: leave the fruit on the tree or hire someone. Some people leave the fruit on the tree, because with the wage you'll pay you won't make any money, but I prefer to give a job to someone, so that he can make a *jornal* for a couple of weeks." As I discovered during my fieldwork, for some of these workers the possibility of "giving a hand" to Bernat and other farmers like him is a crucial source of income.

Yet Bernat is caught in a vicious circle. "I am always behind schedule. I want to put irrigation in all my parcels, but I only have it in three, I don't have time. And this is not a good way to work, you don't enjoy it [*no ho fas a gust*]." The situation sometimes becomes quasi nonsensical. On occasion, the subcontracting company for which Bernat does the refueling calls him to do other temporary tasks in the nuclear plant. Bernat does not want these jobs because they are not well paid (slightly above one thousand euros per month), but fears that if he rejects them he will lose his position in the pecking order for the refueling, on which his household depends. So he takes these jobs, spending the entire salary to hire someone else to farm his land in his absence. The semi-proletarianized situation is a major handicap to developing new approaches or making investments of the kind illustrated by Antoni and Raquel.

We may contrast Bernat's case with Martí's. Martí worked for fifteen years as a truck driver for a cement company. During this time, he was what the locals call a "weekend farmer," helping his father cultivate his land during the weekends and evenings. At the end of the housing bubble, in 2008, he lost his job. Unable to find stable employment, Martí started to make a living by combining the farming of his own land with a variety of temporary jobs, mainly as an agrarian wage laborer (for instance, working for Roger, a childhood friend).

When I first met Martí in 2010, he was receiving the state unemployment subsidy. He saw himself as an unemployed worker and a weekend farmer: "I take any temporary job that I can do, whether in the nuclear plant, or helping in the harvest, or whatever. In the meantime I go *al defora*. I can't really live from it, but at least we have food at home." The last time we spoke, in 2014, his perception of himself had started to shift. He was devoting far more time to tending his own land, both as a result of his father's health problems and of his inability to find a stable, well-paid job. "I was not oriented to agriculture, I was oriented to industry. But what are you to do? My wife is also unemployed, and at least the land feeds us. I made a new vegetable garden and maybe I try getting into organic production." Although Martí insisted that he would take any "industrial job" if he had a chance, he had largely stopped looking for jobs, increasingly seeing himself as a reluctant semi-proletarianized farmer rather than an unemployed industrial worker cum weekend peasant.

The contrast between Bernat and Martí is clear. For Bernat, engaging in multiple sources of income is the only way to be a farmer; for Martí, the relative incapacity to find alternative sources of income forces him to be a farmer. Yet in practice, their livelihoods are quite similar. These two cases in fact show that the borders between the different categories of farmer are quite porous and flexible, and that the inclusion into one or another category depends on a combination of objective conditions, changing livelihood strategies, and subjective attitudes.

Weekend Agriculture: Between Pleasure and Torment

The third category is the murkiest one. It includes a diversity of households often referred to as "weekend farmers." The inhabitants of Fatarella often use a different expression, saying that these people practice agriculture as a "hobby." This expression is based on one of those idiosyncratic syllogisms that illustrate the calculating behavior of farmers: the inhabitants of Terra Alta argue that the harvest "does not pay the *jornal*," therefore, it is a hobby.

Salvador, who works full-time in the nuclear plant and cares for his hazelnut fields, says, "[I]f you make numbers, you just don't do it. You need to see it from a different angle: you eat what you do, you keep yourself healthy, you spend some time with your family, you get some exercise, you keep the territory well arranged, you have some hazelnuts to give away to your friends ... If you make numbers, you better go live elsewhere." Similar logics apply to other self-provisioning activities. In Fatarella, I had a friend who drove about twenty kilometers once a week to collect wild herbs for the two dozen rabbits he raised in his house: "You are what you eat, so you are what is eaten by the things that you eat," he used to tell me. Given the time and the cost of gas involved in the feeding, I once asked whether it would not be cheaper to buy fodder: "Probably, but then why would I raise them? Also, I don't like the rabbits from the store."

Nonetheless, characterizing "weekend agriculture" as a hobby is deceptive in at least four ways. In the first place, it supposes an abundance of alternative sources of wage labor. Second, as Salvador's quote suggests, weekend agriculture fulfills a series of functions—gift-giving, maintaining the landscape, family gathering—that cannot be simply labeled "leisure." Third, there are social sanctions that reinforce this activity. Thus, in the bars it is not rare to hear praise for those who have their fields in very good shape—"combed" [*pentinats*], as they say in Terra Alta—and mockery for those who do not: "I went through X's field this morning, it's full of weeds!" Finally, for some households "weekend agriculture" provides essential, if complementary, revenue.

Miquel has a stable job as a clerk in the municipality, while his wife, Concepció, works in the nuclear plant. They have two teenage daughters, and they are comfortable. Miquel's older brother lives with his parents and is a full-time farmer. Miquel devotes his free afternoon and most weekends to working with his brother (and his father). He even takes two of his four weeks of paid holidays in the winter to work the olive harvest. For Miquel, farming may fairly be described as a hobby: "Some people take their holidays to go do trekking or skiing, don't they? That's not my thing; I don't like getting tired. I like going to the olive [harvest], and I enjoy it more every year. Today, for instance, it was an inclement day, cold, windy, but now I got home after the whole day out there and I feel great. If I could live off of this I would do it right away."

Mercè, by contrast, works as a precariously employed tourist guide and lives with her retired father and her brother, a physiotherapist. Both Mercè and her brother cultivate their land in their spare time, complementing their household's limited and variable income. Like Miquel, both Mercè and her brother say that they would like to live off of agriculture. But whereas Miquel's comfortable economic position allows him to frame his farming engagement in terms of leisure, for Mercè, farming emerges as what the locals call a "calamity": an unpleasant form of labor, a torment. "We've had periods of misery and suffering. If I am working in my land, the fruit that I produce should be rewarded. But it isn't. And my parents suffered a lot, and farming is then seen as suffering, and it is hard to have esteem for it."

Rather than viewing Miquel's pleasure and Mercè's torment as discrete experiences, we can see pleasure and torment as the two contradictory poles around which the subjective experience of the inhabitants of Fatarella constantly pivots. Bernat provides a good example of the ambivalence of subjective attitudes, critically mediated by an objective magnitude, agrarian prices:

> If prices go well you work *a gust* [at ease], and more people will do it. Even the person who does it after the factory shift; if she sees that she can make some money, she will do a better job and put more *cura* and effort. But if you don't make money, then you need to get other jobs and then it is a calamity, you get behind, and you don't work *a gust*. If prices went well, working as a *pagès* is the best: you start early in the morning and you can have a three-hour nap, watch the Tour de France and go back to work and you still make a ten-hour workday, but without rushing [*però tranquil*].

Treballar a gust—the idiom translates as both "working comfortably" and "working willingly"—is linked to *cura* (care, see chapter 3), to the capacity of taking good, reciprocal care of land. Its opposite is *malviure*, literally to "bad-live," an expression locals use to refer to the impossibility properly to farm the land due to lack of time or profitability. Working in these circumstances is a "calamity." With low agricultural prices, locals *malviuen*, using the same expression they would employ if they were working someone else's land.

In the abstract, the difference between the weekend peasant and the semi-proletarian farmer is clear. The former engages in wage labor to strengthen her farming enterprise; the latter engages in farming to supplement her waged existence, either by supplying additional revenue and/or by providing a platform for the fulfillment of social obligations and self-realization. In practice, the difference is far less clear. Although the semi-proletarian farmer tends to be more market-oriented than the weekend peasant, it is a contrast of degree, not kind. Most semi-proletarian farmers devote an important amount of their produce to self-provision: almost all devote certain olive groves to the consumptive needs of their extended families, and many possess small patches of saffron and fruit trees (mostly apples, apricots, peaches, and figs) for that purpose. Conversely, weekend peasants engage in market-oriented agrarian investment. Many have replaced the traditional olive variety of Terra Alta, *farg*, for a variety called *arbequina*, which has offered a better market price in recent decades.

The expression of feelings of moral obligation does constitute a major difference between the two categories. In contrast to her semi-proletarian counterpart, the weekend peasant justifies her activity through a series of morally loaded expressions of obligation that provide a counterpoint to the characterization of this activity as a hobby. She insists she farms out of love for the land, respect for her parents, the guilt that would come with letting the land go to waste [*erma*], or a desire to show her children what it is to be a *pagès*. The weekend peasant frames her activity as one of maintenance or conservation. Thus, it is said that a farmer *porta les terres* [farms, lit. "carries the land"], whereas a weekend farmer merely "maintains" the land [*manté les terres*]. This expression carries a strong moral sanction—it is widely believed that "once you abandon your land, you don't go back to it," an argument confirmed by experience.

Maintaining the land is often framed with a sense of obligation toward the past, but it does contain an important forward-looking dimension: if one ever wants to go back to *portar la terra*, it is necessary to maintain it. Martí's case illustrates this idea, for his current precarious living is afforded by decades of *maintenance*. Similarly, Mercè's brother recently took a leave of absence from his job to make farming his principal activity, starting organic production.

Maintaining the Land or Letting it Go to Waste

Those who do not directly cultivate their land form the last two categories. The feelings of obligation described earlier are crucial in understanding why some households become "sentimental farmers." Through a variety of informal arrangements—which depend on the quality and extent of the land and the commodity crop planted on it—these landowners hire wage laborers to maintain their land and harvest the produce. In exchange, the landowner usually receives a portion of the harvest, which is siphoned into self-provision and the local gift economy. These arrangements, thus, represent a net transfer of non-agrarian rents into agriculture. The classic example that the inhabitants of Fatarella cite is the widower who spends her pension and savings to keep the fields that her husband and/or her parents farmed throughout their lives. The category also includes families with well-paid jobs—such as nuclear workers—who can afford to contract wage labor and do not enjoy or have the time for farming.

Consider Carles, the only son of a strong *casa*; he migrated to Barcelona in the early 1980s. He worked for three decades as a well-paid construction worker, devoting part of his salary and also his wife's (she is also from Fatarella) to pay for the maintenance of his land. During all this time, he kept his house in the village, tastefully remodeled in the 1990s, and visited almost every weekend. Whenever he could, Carles took part in the harvest, together with his sister and her family, who do not live in the village either. "I lost money, sure, but in our family we've always been able to eat our own olive oil, and I used to bring some to my coworkers, and cherries and all that. This is how we've stayed together."

In 2012, Carles was forced into early retirement and moved to live permanently in the village, while his wife stayed in Barcelona: "I always wanted to retire here and take care of my land, which I think is the best way to live. I didn't expect it to go like this, I thought I would retire in five or six years instead of now, but this is how it is. And thank god that I had this, because you know, [when you are unemployed] you start staying at home and watching TV and it gets into your head. But here I go *al defora* and I am fine." He then gave me a few kilos of cherries from his harvest.

The final category describes those who have abandoned their land. It includes a sizable proportion of the population, but a much smaller proportion of the land, for those with the most land have generally tended to maintain it. Around half of the landowners who have abandoned their land do not live in the village, demonstrating the strong correlation between taking care of the land and living in the region. This relationship notwithstanding, it is also remarkable that around a fourth of the nonresident landowners "maintain" their land (at least in part) either by farming it directly on the weekends or by contracting wage labor, often with the intention of spending their retirement years in the village. Abandoned land is often read as a symptom of local decay: loss of population, recessive agriculture, miserable agrarian prices, and sterile households. Nonfarmed land is locally construed as *waste*: the forest eats it and it becomes *erma* [waste]. From the perspective of energy companies what makes the local land waste is its relative lack of market value, whereas from the local perspective what makes the land waste is the fact that it does not reproduce peasant livelihoods. Instead, cultivated land, land that is maintained, is always seen as potentially reproductive.

Land and Livelihoods in Perspective

Agriculture in Fatarella is certainly recessive and precarious. Yet, we should avoid seeing it as marginal, unimportant or uneconomic, for it is crucial for the making of livelihoods. Indeed, although there is abundant evidence that "one cannot live off of agriculture," agriculture plays a critical role in the overall reproduction of the village. For one, agriculture is the main source of income for a third of the village's households and an important income stream for many more. This is a sizable amount, making agriculture probably the main source of revenue in the village. Agriculture may not provide living wages, but alternative wage opportunities are scarce and often poorly paid.

To appreciate the role that agriculture plays for households, let us consider the wage jobs in the village. In my survey, I calculated one hundred seventy-six villagers with full-time jobs. About fifty of these jobs are in industry and ninety in the services sector, with the rest divided between local construction firms and agriculture. Services jobs tend to fall quite evenly into two opposite poles. Well-paid, fairly stable jobs, often in the public sector (nurses, municipal bureaucracy, etc.) coexist with precarious, low paid ones. For instance, twenty women in this category make eight hundred euros a month working for a local cleaning services company. To put that in context, temporary jobs in agriculture are paid between twelve hundred and fifteen hundred a month.

The main source of industrial jobs are the nuclear plants, employing twenty-six people, followed by a couple of small local factories that make furniture, employing less than ten people each. The availability of industrial jobs has dwindled in recent years. A garment factory employing more than fifty women closed in the late 1990s, and the closure of a large tanning factory in Móra and the sharp downsizing of an electro-chemical plant in Flix closed off many opportunities for men. Job scarcity has translated into a new attitude toward temporary jobs in the nuclear plants, historically seen as undesirable; as the locals used to say, elliptically, "too many people end up having to retire too young from there." Today, the nuclear plants have little difficulty meeting their temporary labor demands.

In addition, the economic crisis effectively eliminated job opportunities in the construction sector, both within and outside of the region. Official statistics show that since the outset of the crisis, in 2008, the inhabitants of Terra Alta are increasingly dependent on state benefits (retirement pensions, unemployment benefits, etc.), which in 2013 represented 27 percent of family income (for only 18% in 2007).[17]

Local economic anxiety is not caused as much by the poor state of agriculture, as by the increasing difficulty local's face in finding additional sources of income. Given the lack of jobs, agriculture has become the only activity that offers some flexibility and capacity for self-organization.

Second, despite the feelings of despair that surround agriculture, most locals keep investing in their land. The best example is the recent installation of irrigation systems. When I first arrived in the village in 2010, I was told that no one would pay for irrigation. Two years later, when the water arrived, almost all farming households installed the heavily subsidized irrigation systems in their best parcels.

"It is throwing the money away," most people conceded, "but what are you going to do? Not install it?" Even poor "weekend peasants" such as Mercè and her brother have, against their father's wishes, decided to install irrigation in order to improve production. Like many others, Mercè explains that they have paid for it with the economic compensation received from the wind farm (she was affected by a long ditch that provoked the loss of two harvests). Locals maintain that "if prices ever get decent again" irrigating the fields may make agriculture viable and, if not, installing these systems would make their property more valuable in case they ever want to sell it. In contrast to wind energy development, irrigation is seen as increasing the value of land, which is only an economic asset insofar as it stays in production and is updated.[18]

In the third place, and perhaps most importantly, land provides the support for the internal generation of job opportunities, local solidarities, and rent redistribution. The best example of this mechanism is the hiring of wage labor to "maintain" the land, but there are many other examples, such as the handful of jobs in the local cooperative and the periodic hiring of labor for specialized agrarian tasks such as pruning.

The first time I accompanied Roger's crew on a working day, I realized that the miserable situation of agriculture has led to a redefinition of what is to be a *pagès*. The crew comprised four young men—between twenty-five and thirty-five—with two things in common: they all said they were "school dropouts" and, until recently, they all had been working in the construction of large infrastructure. Iban and Edgar, with whom I spent most of the morning, had been involved in the construction of a high-speed train line and an airport, among other projects. The end of the construction bubble had left them jobless.

While weeding a field of almond trees, they lamented their circumstances: too much work, too little money. "Once I'm done here, in the evening, I'll have to go work in my own fields. You see, this is not life, my wife always complains that I barely see our daughter; this is a torture [*un penar*]," Edgar told me.

The complaints were interspersed with comments that exposed their deep knowledge of the territory. Edgar mentioned that the day before he had eaten a squirrel that he hunted. Iban said he hated hunting, but foraged mushrooms: "When we get to the end of the field, we'll stop for a moment, because in the patch of forest next to it there used to be a place where there were *pebrassos* [a kind of mushroom]. We'll look for them."

At some point, I asked Iban, a member of a landless household, if he found it difficult to adjust to his new life situation. He replied:

> Until two years ago I was working in civil construction, in Barcelona, Madrid, Girona … doing highways, airports, high-speed train lines, all this stuff, as many other people of my age from here in Fatarella. I was making good money. But then the crisis came and there was no more work, so I came back and now I work doing all kinds of things. I mean, I am *pagès*, so I can work in all sorts of things. In the last months I've worked plumbing, given a hand

in construction to people fixing their houses, and mostly pruned olive and almond trees, mostly for people who want to have their land *arreglada* [well-arranged, neat, dressed up].

I asked him if he did those jobs legally, with a contract and reported income. He blinked incredulously. "Of course not.... But what I was telling you: I've been lucky. Many friends of mine haven't found so many things, so then they work in their land [*van al defora, treballant a lo seu*]. We are *pagesos*, see? We work in whatever we need to. We learn how to do all sorts of things, and we have no problems in working the land. It is not like the people in the city. Here you can never be fully jobless. You are never in total misery."

For Iban and his generation, being a *pagès* means something different than for Daniel. Pluriactivity is not exceptional for them; for Daniel, being a *pagès* and making a living *off the land* were synonymous. Iban believes being a *pagès* is about constructing a livelihood in which the land is one among several other supports that allow for making a living *in the land*, in the village. Being a *pagès* indexes a resourcefulness that makes that living possible and even desirable. By defining himself as a *pagès*, Iban was claiming his dignity, connecting his living with a long struggle for autonomy, in which torment and pleasure, survival and resistance, precariousness and resourcefulness conjoin a coherent practical consciousness that turns nostalgia into an actual resource. Dignity is as connected to the process of eking out a livelihood as it is inseparable from the search for coherence and meaning. Maintaining the land is critical for this endeavor.

On Blood and Stone: Notes on the Art of *Apanyar Pedra*

In 2007, three landowners affected by the construction of Terraca's electric line attached to their *acta de expropiación* a letter intended to express the feelings of defenselessness and disrespect provoked by the expropriation process. In a key paragraph, the letter glosses the "moral and intangible" value of their property and, by extension, the agricultural land of the county:

> The affected area should be valued not only as a certain amount of square meters, but as a land that has been built and modeled over many generations. Its moral and immaterial value does not have a price. Our land is as important as the Ebro river or as the Sagrada Família temple. No one wants to undo the Sagrada Família to build a parking lot.... They have humiliated us, telling us that they were doing us a favor [by] giving us money for a land that had no value. Forgetting that our land is much more than just square meters, forgetting that this land has been made *a cop de sang*.... People's dignity should be respected.

To these landowners, wind energy companies devalued the labor accumulated in the land—a labor that has made the land and its inhabitants what they are. To express the work of generations, the people of Fatarella use a blunt idiom:

a cop de sang. The phrase plays on the expression *a cop d'aixada* ("tilling"). By substituting blood (*sang*) for hoe (*aixada*), the idiom conveys the idea that the land has been built through the toil of generations and that such toil is embedded in the land. In opposition to the wind developers' vision of land and local practices as a wasteland, the local population cultivates a different vision of their agrarian space: it is understood as the outcome of a series of relations and practices. The struggle for the construction of a peasant order is largely played out in this spatial domain. Land and livelihoods are intricately cobbled together. If practices and relationships are what make land valuable, this value can only be retained, preserved, if those practices and relationships are *maintained*, reproduced.

It is no coincidence that Remigio, president of *Ventdetots*, a local organization that opposed the way that wind farms were implemented, has a small company devoted to drystone construction. Like no other, this vernacular building form encapsulates the work of generations, materialized in the making of place, and it is dear to the heart of the locals. Drystone construction populates the landscape with walls, dwellings, terraces, stairs, wells, field entrances, cisterns, pathways, coops, and pools.[19] These elements epitomize the locals' relationship to land as the secular, constant task of conditioning, modeling, and rearranging their environment. Drystone architecture is made of toil and pride, despair and resistance.

An Art of Doing

Locals rarely call it "drystone architecture." Instead, Southern Catalans use the idioms *apanyar pedra* (fixing or arranging stones) or *fer marges* (making walls). Their practice is a prime example of what Michel de Certeau calls an *art of doing*, a form of practical knowledge, in which thought and practice form an indissoluble unity: "A way of thinking invested of a way of acting, an art of combination which cannot be dissociated from an art of using."[20] Indeed, *apanyar pedra* is contextual (its field of action has no autonomy), combinatorial (building from a repertoire of possibilities provided by experience), tactical (structured through time, never establishing a finished order), and popular (it does not constitute alongside it a field of abstract knowledge controllable by experts).

Over generations, Southern Catalans have modeled and shaped their land and made it the basis of their livelihood. This is true in a literal sense: most of the arable land of Southern Catalonia consists of terraces sustained by stone walls. Even the exiguous plains that did not require terracing have been made arable only through the constant removal of stones, cleared and collected for wall making. Drystone architecture is essentially democratic: it uses free, readily available elements and tools that every *pagès* has at home. It is labor-intensive and requires a great deal of skill, codified in some technical vocabulary carefully distinguishing, for instance, different kinds of constructions, building actions (*eixobrir, agermanar, fer rentador,* etc.) and stones (*fiter, falleva, galga, reble*, etc.). Although only a handful of *mestres* (masters) in the county, including Remigio, know how to build and repair the most sophisticated architectural elements (such as cisterns or wells), *apanyar pedra* never gave place to a separate trade: all *pagesos* are "stone arrangers."

This contextual knowledge cannot be abstracted from its social and topographic bearings and thus converted into a bounded field of technical expertise. *Apanyar pedra* may be seen as a prime example of the combinatorial praxis that Claude Lévi-Strauss called *bricolage*.[21] Our drystone *bricoleur* deploys a flexible, but limited number of techniques and rules (for instance, one stone must always rest above two stones) and makes use of what is at hand: if no big stones can be found nearby, he will use smaller stones; if a pine tree is growing at the base of a wall, he will not cut the tree but rather reshape the wall to accommodate the pine. Although the drystone *bricoleur* does, on occasion, use a hammer to shape the stone, his main task consists of identifying the possibilities in what he has at hand to create a provisional synthesis, arranging the stones so that they "link"—an action for which the locals use the word *agermanar*, literally "fraternize."

Helping a drystone *bricoleur* can easily make you feel stupid: where one sees a bunch of rocks, the *bricoleur* identifies the elements of a tactical order on the go. After observing and aiding others, I told Bernat that I found *apanyar pedra* to be quite a difficult task. He responded dismissively: "It's just about being careful. You look at the stones and then see which one you have to put on. It's easy, it's all about looking and finding. And those who know a lot don't even need to look; the stones go to their hands on their own."

Fixing Stones, Making Livelihoods

The art of *apanyar pedra* does more than mold the territory. It discloses the reciprocal connection between how locals take care of their land and how they organize their livelihoods, revealing what Gramsci would call a "conception of the world" grounded on material practice.[22] Indeed, *apanyar pedra* is intimately connected with local notions of order. As Remigio explained:

> This last summer, in a field that belongs to my wife, I fixed a couple of stone walls. She said that I shouldn't do it, that I work too much, but I did it. Look, it is a little thing, but the field is a different thing with it. You put the new stone pieces, and you observe them and you notice the human hand. And you say: look at it, all the walls are well done. The parcel is the same one, it has not been moved, it hasn't switched into a new position, it's not bigger or smaller, of course not, but it is tidier [*més endreçada*]. It's like a house where no one swept the entry, it's the same if you don't keep the stone walls. You know, you go to a field and you see that everything is in a correct place, the box with the apples here, the tractor there.... In a word: it is not disorganized [*desendreçat*]. This is what a field with the stone walls in good shape is, a well ordered field [*una finca ben endreçada*], nothing more—but nothing less—than that.

The term *endreçar* that Remigio uses gives us a good clue about the notion of order that is at the basis of the local relationship with nature and of local livelihoods. He does not use the more common verb *ordenar*, which, as in the English, has a dual meaning of the arrangement of elements according to a given pattern and of authoritative command. *Ordenar* thus presupposes a synoptic view, a finished order that will be executed on the world, very much as in legal codes. For its part, *endreçar* means to arrange, to tidy up, and to organize. It shares a root with such English words as "dress," "address," or "redress," which derive from the French term *drecier*, and it is equivalent to the Catalan *dreçar* (which means to arrange, to straighten, or to put upright), derived from *dret*—analogous to the French term *droit*—meaning upright and correct. As these etymological connections suggest, *endreçar* evokes an order that is never complete. It is a permanent task of giving direction, an order diachronically produced through praxis, orienting toward a certain purpose a series of elements that are not of one's own choosing. It is the form of order that emerges out of *cura* (chapter 3), out of the continual action of *taking care* of the environment.

Behind the praxis of *apanyar pedra*, we may also discern a local understanding of nature. The inhabitants of Fatarella often emphasize that *apanyar pedra* has made their territory productive, yet *apanyar pedra* and farming are not seen as the origin of this productivity, for local knowledge conceives of the natural world as inherently productive. Referring to someone who has stopped cultivating a field, people will say: "He has *let* the field become forest." Nature produces on its own, but this productive strength is wild, disordered; through *cura*, *pagesos* turn nature's wild productivity into fertile lands and livelihoods. Nature, in fact, even provides a seemingly inexhaustible amount of stones—no matter how clean the field is, new stones always seem to pop up. Yet this unproductive, disordered element—"stones are the enemy of the *pagès*," locals say—may be converted into the framework of a productive landscape.

This understanding of farming as a task—involving both toil and pleasure—of reproducing a temporary, manageable order extends beyond agriculture and the relationship with nature. Organizing one's life, making a living, is also seen as a task of arranging disparate elements into harmonic ensembles, finding the place for each thing—as the Catalan saying goes: "Each thing in its place, and a place for each thing"—and in so doing making the place for yourself. Southern Catalans' re-patterning extends to the entirety of cobbling together livelihoods through multiple sources of income, embedding different activities and relationships into a meaningful whole. Farming, arranging stone, rearranging the landscape, and making a living become inseparable, a chain of activities that articulates the historical praxis of the inhabitants of Fatarella. For the local inhabitants, living in Fatarella is the product of a long, ongoing historical struggle consisting of rearranging a series of elements that are not of their own choosing into coherent sets. These elements may be stones and dry land, but also scarce job opportunities and semi-proletarianized livelihoods.

The productive yet directionless character of nature is often used to justify another idea: "Once you abandon the land you cannot get back to it. This land that we have here is very wild [*feresta*], all at once it gets full of weeds [*malesa*] and stones." The task of *cura* and *endreça* must be constantly performed, and if you disregard it, it quickly becomes too late; the order is too fragile, it can only be sustained if *pagesos* continually reiterate the work of past generations, thus actualizing its value.

Again, *apanyar pedra* provides a basic metaphor for this understanding. Indeed, locals rarely build new walls and other drystone constructions. Most often they simply combat the deleterious effects of time, fixing and restoring venerable structures. If you let a stone wall crumble, it is unlikely that you will rebuild it—it is too late, it is too much work. This idea applies to the whole notion of being a peasant. Experience shows that those who abandon their fields never return. If the land is abandoned, it will leave the community, it will escape from the circuits through which locals organize their livelihood.

The purpose of drystone architecture is not only to make the territory productive, but to make it a *locus amoenus*: a pleasant place, a place for joy and comfort. Aesthetic considerations are central to the art of *apanyar pedra*. Making the fields into a *locus amoenus* is seen as the best way to make it a space of sociability and have the whole family involved, as Bernat explains: "If the place is well *endreçat*, if you feel at ease [*si t'hi trobes a gust*], then it is more likely that all the family wants to come and give a hand and enjoy." The care and passion locals devote to their *masos* is the best example of the making of the *defora* as a space of joy and sociability.[23] Nowadays, *apanyar pedra* plays a minimal part in the conditioning of *masos*, but the operational logic is the same. *Masos* are dressed up by recycling what others see as waste: war rubble and old farming utensils are used as decoration, chairs gleaned from an old night club are reused, and industrial materials obtained for free are put to new uses.

Creating a *locus amoenus* requires faith in its nurturing, reproductive effects. The farmers of Terra Alta believe that if you spend enough time in the fields as a child and a youngster, you will never be able fully to abandon them. In contrast, "if you don't know it, if you're not raised as a *pagès*, you will never get into it." There is also the well-founded idea that those who leave may only maintain a link with the village if the land is well organized, if the fields are in production, if they can keep engaged in exchanges of agrarian produce. The connection is critical to maintain a transfer of rents from the city to the countryside through, for example, urban dwellers' extensive renovations and rehabilitations of houses they own in the village (as we saw in chapter 3).

Order as *endreça* does not only apply to the making of livelihoods, but to the whole social universe of the village. As with nature, society is seen as permanently productive, yet these energies are potentially disorganized. The war represents the epitome of social and natural disorder: a moment when neighbors turned against each other and the land of the village became useless due to carelessness and the detritus of the conflict. The task of fixing the *defora*, taking care of it, is widely understood as a task with collective benefits: "The municipal territory is everyone's, we all see it." Making the *defora* enjoyable, pleasant, requires some extra time, a certain surplus of pleasure and generosity, energy devoted not to production but to the collective endeavor of modeling the territory.

Analogously, keeping social order, maintaining *convivència*, requires some extra work: if you keep to your own business, if you just do the minimum to reproduce your livelihood, *convivència* is likely to erode. Solidarity and autonomy must go hand in hand. This is why the wind farms—which involve an alteration of the landscape with purely private benefits—and the difficulty to make a living are seen as threats to social life, for they erode the capacity of the locals to *take care* of their socio-natural universe, to give order and direction to it. *Convivència* requires some spare time, some time for leisure, some reciprocity, some generosity. As we saw in chapter 6, the land leases for the wind turbines made quite evident that most residents did not have—or did not want to use—that margin of maneuver. Self-interest trumped reproduction. In fact, this was far more damning than treating Fatarella *as* a wasteland, for it suggested that perhaps it *was* a wasteland.

Dignity and Indignation

As we have seen throughout this book, confronted with a series of energy projects that threatened their livelihoods, Southern Catalans have reacted with indignation and assertions of dignity. How are we to understand these moral exhortations around dignity?

As an affirmation of intrinsic worth, human dignity has become the axiomatic basis of human rights, a cornerstone of international liberal law. Nonetheless, dignity puzzles political philosophers and legal scholars.[24] Although I do not aim to undertake a philosophical inquiry into dignity, it is worth observing that the origin of this puzzlement is the indeterminacy of dignity, a feature it shares with the notion of waste. On the one hand, dignity is one of those concepts that philosophers call "thick," for it carries both descriptive and evaluative content.[25] It cannot be examined outside of social practice and specific cultural constructions, frustrating any attempt at a purely conceptual discussion.

On the other hand, scholastic arguments over whether the notion is rooted in hierarchical or egalitarian conceptions of society stress its *political* ambiguity. Dignity may point toward conservative, antidemocratic projects as well as radical, emancipatory ones, philosopher Stephen Riley notes: "Human dignity ... contributes to the disruption of a simple, hegemonic, idea of sovereignty. It encourages ideas of popular sovereignty; it also invokes hierarchical norms that mesh with the colonial ideology of civilizing and subjugating."[26] Indeed, it is my contention that this disruptive capacity is the key to understanding the notion of dignity and the moral exhortations that surround it.

In *Natural Law and Human Dignity*, Ernst Bloch aims to capture the radical, emancipatory, republican spirit of dignity. Dignity, for Bloch, is rooted in natural law and lies outside of a political economic order in which legal equality and freedom coexist with economic exploitation and political subjugation. It points to the simple, fundamental right "not to be treated as scum."[27] And that is alien to capital's law of value, for it emerges out of an understanding of the economy as livelihood. Consistent with Kant's critical Enlightenment—"Everything has either a price or a dignity," wrote the sage of Konigsberg—the search for dignity is a fight against degradation that can neither be reduced to, nor separated from, the struggle against economic exploitation.[28]

Even the young Marx invoked dignity, the indeterminate *humanum*, to denounce proletarian alienation: "This class is, within depravity, an indignation against this depravity, an *indignation* necessarily aroused in this class by the contradiction between its human *nature* and life-situation, which is a blatant, outright and all-embracing denial of that very nature."[29] As Werner Bonefeld argues, Marx's later critique of political economy, showing forms of capital to be perverted social relations—concealing exploitation and dispossession under a veneer of liberty and legal isonomy—can be understood as an analysis of human dignity as *dignity denied*: human purposeful practice alienated from its conditions in the form of capital.[30] Positing value as an alien, illegitimate element, dignity emerges as the "other" of capital's *constitutive premise*, the separation of people from their means of reproduction.[31]

Thus, I argue that waste and dignity operate in opposite directions. If, as Gidwani argues, waste is the "other" of value, dignity, I would add, is the *other* "other" of value. Whereas waste indicates worthlessness, dignity, at its bare minimum, indicates, at least since Kant, intrinsic worth. If waste is the indeterminate negative of value, the retrograde residuum that capital needs to conquer, dignity is, Bonefeld and Psychopedis write, "the polar opposite to commodity exchange relations where every human being has its price ... Dignity cannot be sold, quantified or conferred."[32] Whereas waste opens a possibility for the expansion of capital's law of value by mobilizing new resources (means for profit), dignity points to human beings as ends, subjects of their own social world. This is how I propose to understand Southern Catalan calls for dignity, as a struggle to reproduce a practical sense in front of imposed order, a struggle, to paraphrase Gudeman, against "being modeled."[33]

Southern Catalans use the notion of dignity to disrupt the law of value, which condemns them to dependence and marginality. Dignity does not emerge as some sort of ideal causal factor but as the expression of a historical framework of material experience, as the condensation—what Raymond Williams called a "practical consciousness"—of a structure of relations and practices.[34] The demand of dignity operates as a refusal of the political economic state of affairs that has governed the region since the 1970s, a process that I described in chapter 1. Because it is a demand for dignity denied, this demand always has a double valence, expressed in two different moments that we could call *indignation* and the *assertion of dignity*.

As Daniel Bensaïd points out: "We still have the irreducible force of indignation, which is the exact contrary of resignation. Even when one ignores the justice of what is just, there is still the dignity of indignation and unconditional rejection of injustice. Indignation is a beginning. A way of rising up and getting going."[35] The shared, visceral affect of indignation emerges as a negation of what Bourdieu, drawing on Spinoza, called *obsequium*: "unconditional respect for the principles of the established order."[36] Indignation is a rejection to accepting the uneven ecological and economic distribution that affects Southern Catalonia, to remaining silent when you are told that your lands are worthless, or to becoming pronuclear just because you need the job. Indignation appears as the assertion of a resistant subjectivity stubbornly struggling against the denial of dignity, as we may appreciate in the words of Mercè:

> Everyone goes to work elsewhere, to the city and the nuclear power plant and all this. Here we should all live from agriculture; we all come from [farming] here. I have a right to be a peasant, to make a living from agriculture, because I come from a peasant background, my roots are peasant, and my father is a peasant, and after the war he worked this land with his hands when it was all devastated. We should all live like that. We have resources: we have high quality products and we have the capacity to preserve and transform them. But I need to be paid "dignily" [*dignament*] in order to live with dignity.

Mercè, who is precariously employed as a part-time tourist guide, is a *pagesa*, yet she cannot be a *pagesa* even if she is condemned to be a part-time *pagesa* to make ends meet. Mercè expresses what Gavin Smith calls "resistant subjectivity," tracing a continuity between this subjectivity and that of the resistance fighter: "Resistants are in many ways the political expression of perpetual insecurity and risk who participate in a spasmodic and often concealed social life fighting revolutions that are always only provisionally won, returning to the present from perpetually unrealized futures."[37] By asserting her peasant identity, Mercè is continuing the stubborn struggle for autonomy that has historically characterized the Southern Catalan peasantry.

For most people in Terra Alta, taking care of the land—and even staying on the land—is both a form of livelihood and a form of resistance. This livelihood can only be maintained through the reproduction of a resistant subjectivity: maintaining this subjectivity and maintaining the land are hardly distinguishable. The reproduction of this resistance is pivotal for the reproduction of certain practices and relationships that make living in Southern Catalonia possible and worthwhile. The alternative would be resignation, reversing *indignatio* into *obsequium*—deference, obedience. Indignation means refusing to accept the energy company as the *amo*, to be dependent on it, to participate in its moral order. This is why the recurrent expression "I wish I had three turbines" is such a symbol of defeat, for it signals the replacement of a resistant subjectivity for a deferent, or at least passive, one that accepts the moral order of the wind developer, in which lands are waste precisely because redundant people and ineffectual practices inhabit them.

Mercè's indignation demonstrates what Ernst Bloch, as we saw in chapter 1, calls nonsynchronism, the expression of an uncompleted past, the unfulfilled project to "live from the land." That impossibility is expressed and experienced as a matter of agrarian prices, translating into an iterated demand for "fair prices" (*preus justos* or *preus dignes*) that reveals what EP Thompson called a "popular moral economy."[38] Indeed, fair prices are very clearly defined in Southern Catalonia as prices that allow for making a good living. "Good living" is also quite well defined: that which allows one to take care of the land and oneself, to work comfortably, *a gust* (with pleasure, at ease), without suffering. The argument is self-referential, certainly, and consciously so. This is precisely the point: being able to control your time and space. Dignity should not be confused with resistance or agency; it constitutes a precondition and latent possibility for them. As Nick Bromell argues, "dignity lies dormant ... until it is recognized or denied by another."[39] It is the denial of dignity that articulates the relationship between the two moments of the demand for dignity: indignation and the assertion of dignity. Or as locals used to tell me, "By expressing your indignation, you show that you have dignity."

This logic is powerfully captured in the Spanish (and Catalan) verb *reivindicar*: to vehemently (re)claim something that one considers to belong to oneself. This *revindication* of dignity takes two main forms: asserting the value of the area—including both land and people and, most crucially, the relationship between the two—and the search for autonomy and *autogestió*, that is to say, the capacity to be your own *amo*. The first claim asserts itself against the construction of the region as a wasteland, the second against the forces that place local inhabitants in a position of dependence and degradation. These two demands emerge as efforts to preserve relations that already exist. The demand of dignity is an attempt to preserve—but also to construct and perform—certain possessions, a struggle against dispossession: being able to make a living and stay put, preserving the value of land, maintaining local networks of solidarity, preserving some control over the labor process and household reproduction.

The *pagès'* demand for dignity is part of a struggle to establish an order of elements conducive to the reproduction of the community and its individuals in front of a series of forces that introduce disorder, that construct the community and its people as waste. Southern Catalans do not wish for a "return" to a rural arcadia, but to govern themselves, to undo the division between country and city that lies at the center of capitalist development and its production of subordination. Their opposition to wind farm projects is informed by the memory of the antinuclear struggle, for wind energy development has reproduced a centralized, concentrated model dominated by the "*amos* of energy" that is antithetical to the renewable future people like Mercè had imagined during the antinuclear struggle.

From this perspective, the demand of dignity may be nonsynchronous, but it is not oriented toward the past. Dignity is oriented toward the present: not so much the effect of the coexistence of distinct temporalities or the expression of an uncompleted past—even less of a universe of non-commodified relations—but rather the assertion of values and practices that are alien to the abstractions of capital and its law of value. Demands for dignity are not blind to political economic structures and constraints; the local population is all too aware of them. Dignity expresses a desire—even a need, I would argue—to think of oneself outside of those structures and the immediate contradictions that they produce. It is an assertion that locals are not only semi-proletarian, but also resistant peasants that try to make sense of their lives. This resistant subjectivity connects indignation with a drive to mobilize collectively, for, as Josep Maria Antentas argues: "Struggling requires not only malaise and indignation; rather one must also believe in the usefulness of collective action, in the fact that winning is possible and that not everything is lost before the beginning."[40]

Visions of the Future

The demand of dignity represents an effort to control local time and space, thus being intimately intertwined with wind energy development. As Werner Krauss argued in the case of the German North Sea, the position that local dwellers take on wind energy development must be understood as a struggle to "maintain, shape and administer the environment they live in."[41] In this respect, the demand of dignity is not only oriented to the past and the present, but also toward the future.

In Southern Catalonia, many see wind energy development as an impediment to the creation of future horizons, a marker that the region will not be able to reverse a historical trend of peripheralization and impoverishment. In this concluding section, I will describe three different life projects that have been noticeably affected by the installation of wind farms. As I show, these life projects can be understood as efforts to "maintain, shape and administer" the local environment, to create a place conducive to making fulfilled livelihoods. The frustrations of many Southern Catalans highlight the unrealized vision of a noncentralized energy system that could erode the country-city divide and the power of electricity utilities. As the stories here and elsewhere in this book show, dignity is not just the cultural code through which economic and ecological imbalances are locally understood; it also possesses social and political possibilities for the analysis of energy transitions and the construction of energy futures.

A Peasant Landscape to Return to

Remigio, the president of *Ventdetots*, connects wind farm development with the destruction of an old ecological order. His opposition to the local wind farm should be understood as the defense of a peasant landscape modeled through generations:

> We spent the last two weeks reconstructing pathways in the Serra of Llaberia, old peasant pathways that now are being restored for the use of trekkers. How can I not be angry with the *molins* people if they trample on the municipal territory? I am from Fatarella, I've been born here and I love it, and for me this means dignity. If they come here and try to change everything I have to rebel. I know they will do it, and they did it, but I don't agree, this is the minimum I can say. Why? Because I have dignity, I am not sure I am clear.... They say that the wind farms are ecological, and they are, but they aren't. Ecological means making the best out of something [*ben aprofitar*], not destroying, going around serenely [*tranquil*]. Forty years ago all the villages were ecological: no insecticides, no residues, we had chicken and rabbits and a pig, the leftovers for the dog, the garbage for the pig. Now everything goes to the trash.... We had a chain, and we don't have it anymore, progress... Look, I don't know much, but I feel that we haven't been very good at applying this progress, we have been depopulating the villages, everyone leaves, agriculture is sinking, and it's been like this for decades. I have a tractor that I bought thirty-five years ago for 675,000 pesetas [roughly, 4,000 Euros] with a single hazelnut harvest. Nowadays, the tractor costs 30,000 Euros, and for the hazelnuts I get 4,000 Euros.

The critique of the centralized model of wind farm development is obvious, for Remigio connects it with the dynamics that have led to the depopulation of Southern Catalonia and other rural territories targeted for wind energy development. Yet, he also suggests that this peasant space can be the basis of an ordered management of the territory, as he demonstrates through his own paid work as a drystone mason.

Remigio's opposition to wind farm development as a position oriented toward the future became especially clear when he told me the main reason he embarked on his crusade: his only child, Elisenda, asked him to. Elisenda is in her early thirties, works as a nursery teacher, and is married to Ricard, a civil servant. Although they are both from Fatarella, they live in Tortosa, Southern Catalonia's main town. They visit the village every weekend and dream of being able to make their living there. Elisenda explained:

> For us this is quality of life: your family, your land, your roots, your friend-ships, the organizations of the village, everything. We are members of many local organizations, I mean, our life is here, but our jobs are elsewhere. We love the village. Ricard is studying history and thinking of doing a book about the village. I love going to the fields, for me it is all about the land, going to the fields, feeling that smell ... that reminds me of my childhood, when I went to the fields with my father, my mother and my granny. I need it. It is my way of disconnecting, going to the fields, I want to live [on] the land, feel the land. You know, the village is ok, it has nice places and everyone is different, and generally I like it, but the thing is the land.

Elisenda's opposition to wind farms did not emerge, at least not solely, out of a nostalgic connection with the smell of her fields. Rather, it was gestated during the Southern Revolt, a movement that signaled hope and a possible economic future for the region based on the locally controlled management of endogenous resources. However, the wind farms and the other energy projects that keep emerging make the couple feel that it may not be worth coming back. Ricard and Elisenda are doubly disillusioned at seeing that Fatarella is being converted into a wasteland and at seeing that their co-villagers accepted the conditions offered by the developer. They see that farming is lost at the pace village population is lost. Wind farms feel like another, perhaps final, nail in the coffin:

> I don't think we have any dignity, to be honest. In my household, we had a turbine, and we said no: expropriate. I will never take this money. It was a matter of personal dignity. And everyone comes to me saying that I was too radical. Fine. If someone else feels that they are worthy [*dignes*] of a turbine at home, like in Iberdrola's ad (see chapter 6), fine with me. But this is because people here are bruised [*tocada*]. I feel like crying when I see the turbines. But it's always the same people fighting, just a handful, the rest don't seem to care, they feel there is nothing to do, they say: "One day they'll even take the wages away." And people elsewhere don't think about the other people: people just want to turn on the light, they don't question whether that is affecting some-one else, if someone else is suffering. The situation here is dire, and the wind farms just make it worse. It has degraded our hope.

Elisenda and Ricard are key actors in the future of the county. Highly edu-cated and passionate, deeply involved, these are the kinds of people lagging rural territories need. But when Elisenda looks at the turbines, she does not see renew-ability but rather an epitaph, the marker that there is no hope for them in the area.

A Terroir to Live in

In 2010, Gas Natural and Alstom Wind won a public bid to build three wind farms between the municipalities of Vilalba and Batea, in the heart of Terra Alta's wine country. In 2012, Gas Natural discreetly began to establish local contacts through Eva, who, after the sale of Obrisa, had no other option than to work for a major electricity utility if she wanted to stay in the sector: "When I saw that Gas Natural won the bid, and seeing that the sector was in a bad place with all the talk about the subsidies and all that, I sent my CV to Gas Natural."

These wind farms would add to the two already installed in Vilalba, a village that gets 40 percent of its municipal budget from the wind farms. Indeed, people I spoke with in the wind sector used Vilalba as an example of the benefits that a collaborative relationship between the wind energy sector and small municipalities can bring to both parties. Yet, while it is true that Vilalba's municipal government and, to a lesser extent, its population have been supportive of wind farm development, it is also worth pointing out one of the first things that Ramon, Vilalba's mayor, told me when I asked him about this latest project:

> I personally think we already have enough *molins* here, but we had no say in this project, it was shoved down our throat. We've already done our part. Our contribution to Catalan society in terms of generating energy has been more than fulfilled. I am not against *molins*, but we should not go too far. It brings money to the municipality, but other than that it generates almost no jobs. And the visual impact is substantial. I am a farmer and I don't see them anymore, but you know, you look around and they're there, and you hear them. And some people do see them and they don't like them and maybe these people never come back again.

Although the initial contacts to build these three wind farms were carried out with great discretion, I learned about them through some landowners who regarded the development with reluctance. The people Ramon had in mind—the ones who might be pushed out by wind energy—are relative newcomers like Luca and Judith. About a decade ago, the two enologists were in their mid-twenties and dreaming about their life project. Having worked for a few years in several wine cellars in Northern Italy, where Luca is from, the couple decided that they wanted to create their own cellar. Although they possessed certain knowledge of Southern Catalonia—Judith is from a nearby region—they wanted to take some time to choose the right location. So, in 2005, they both found jobs as enologists in Priorat and spent two years exploring the whole region in search of the right location for their cellar. Priorat seemed like the obvious choice, but it was expensive and they fancied the idea of an area that was less crowded and offered more open possibilities. In 2007, they found it: twenty hectares in Terra Alta's wine country.

Luca and Judith bought the parcel of land and planted vineyards on twelve of the hectares. They built a cellar, for which they imported Italian wine processing technology, and constructed their domestic unit above their small enterprise.

They planned to have some farm animals and add some crops like almond and olive trees: "Our idea is not only to make wine—good, commonsensical wine—but also to build a complete, closed agrarian system, that would then be linked to complementary tourist activities with wine-making at its center."

They even tried to install solar panels and a small wind turbine to make their business energetically self-sufficient, but gave up due to the bureaucratic burdens and the high cost of connecting to the electricity grid. In 2011, they had their first harvest and started bottling their first wine. The labels bore the cellar's trilingual motto—*Interpretando il terroir* [Interpreting the terroir]—a clear statement of the value they give to, and their commitment toward, the territory: "Our wine is made down there [in the vineyards], not up here in the cellar. Up here we just interpret what we did down there. We think of ourselves as integrated in an ecosystem."

In 2012, Luca and Judith discovered a wind meter in a neighboring field. It was like the sounding of an alarm. The couple went to the municipality of Vilalba and found out about the wind farm projects. They were informed that one turbine would likely be located in their fields. To them, it seemed that the wind projects threatened their life project. Their vision was highly influenced by the platforms' arguments at the turn of the century and the defense of an alternative economic development project:

> Do you think this is sustainable? Integrated into the territory? With two of these turbines you could fulfill the electrical demand of all of Terra Alta, so let's put half a dozen, or a dozen, and be supportive with the rest of the country, but we now have hundreds, literally. We are totally ok with wind energy, harnessing renewable resources .... of course; it's the most beautiful thing! But this is not the way, with no information whatsoever and with no consideration toward the territory, as if it were worth nothing. This is not an aesthetic issue; this is a violation of the territory. The problem is speculation and overcrowding, a total lack of territorial planning. They know they're destroying the region, and take advantage of the low income and level of education of the population. They modify a territory on account of its future, they destroy its agrarian potential, they haven't considered that the only vocation of Terra Alta is agriculture and the tourist activities more or less associated with it.

I asked Luca and Judith what would they do if the wind farm materialized. Perhaps naively, they didn't think it would happen. They didn't think it made sense.

"But what if?" I insisted.

"Well, we don't know, we try not to think about that, but if it happens, surrounded by all these turbines, some of them in our property ... I guess we'll go somewhere else. I mean, we aim for integration with the territory, and do you think I can integrate with the turbines?"

In 2012, the three wind farms seemed like a remote possibility, but with every successive conversation we had, despite their complaints and protestations, it became increasingly evident that the installations would probably become reality. In 2015, Gas Natural announced it would start construction in 2017. As a result, the last time we spoke Luca and Judith told me that they were seriously considering moving their life project elsewhere.

In the summer of 2017, Terra Alta's wine regulatory council did for the first time get involved in the matter. It issued a manifesto against wind farm over-crowding and against the installation of the three projected wind farms, arguing that they "imperil[ed] the economic and social viability of many enterprises." And it continued: "The existing wind farms have not brought anything positive and real to the territory.... Wind farm overcrowding may destroy a source of real wealth as well as future investment opportunities by new companies and dwellers that search for authenticity. Wind farms have a burdening and disillusioning effect toward tourists. In Terra Alta, landscape is culture, and the wind farms endanger our cultural heritage, including the [spaces of the] Battle of the Ebro."[42] In spite of these efforts, in the fall of 2017 some landowners started receiving the first official letters notifying them of how the wind farms would affect their property.

Luca and Judith are the kind of newcomers that any rural territory desires: they buy barren land and make it productive, they transfer technology, they are young, they make a strong investment and a lifelong commitment to the region, they have access to markets, and they are invested in "adding value" to the agricultural produce and the land. But they are considering leaving if the wind farms are built. To read their decision as a simple resistance to having wind farms in their back yard would miss a crucial point: the place is not being made for them. The installation of wind farms in Terra Alta signals to investors that Terra Alta is not a *terroir*, but a wasteland, a place where the things that others do not want are installed, where the resource is extracted, while profits and electricity go somewhere else.

The relationship between Luca and Judith's *terroir* and the peasant space I have described is complex. The peasant space defended by persons like Remigio is rooted in the past, in the blood of their ancestors, the memories of being raised up in the fields, the relationships of solidarity and cooperation built between *cases*. Judith and Luca's *terroir* is an assertion of a future value: theirs is the vision of Priorat's winemakers, for whom the *terroir* is the centerpiece of their strategy to create a market niche and add value to their product. The *terroir* is a latent possibility preserved through peasant resistance; the peasant space is a validation of the *terroir*.

There are doubters in Terra Alta and elsewhere in Southern Catalonia who have little faith in the economic project the *terroir* embodies. They have a different vision of what future wealth should be and how local livelihoods could be reproduced. Gonçal, the small contractor and weekend peasant who in 2011 became the mayor of Fatarella, is one of them. He would like to see new investment, new roads, new factories, and, most importantly, new jobs, even as he bitterly admits that wind farms have done little of this: "I mean, if they put the turbines here, why don't they put the [wind turbine] assembly plant here? With all the electricity we can provide.... They have had no interest in creating wealth for this territory. They have just been interested in exploiting the wealth of this territory: the wind, the mountains, the fields." The extractive, centralized character of wind energy development has emerged without any concomitant vision of development for the locals. For the people of Terra Alta, their place is not to be a *terroir* or a modern industrial hub, but rather a constantly needy wasteland.

A Place to Resist

I was absorbed looking at my notes in the *casal* of Fatarella when I heard someone say, "So, what I heard is true: you're back."

It was Iban, who I had not seen since our encounter in the hazelnut fields more than a year earlier. He had big news: "Now I have a *mas*, I am a *masover* [*mas* tenant]," he laughed. "You should come check it out, bring your family." He looked happy as I offered my congratulations and asked how it had come about:

> I was tired of going up and down with the car to find miserable jobs in construction. And I like this thing of the *mas*. I have a big vegetable garden, and animals, and I just planted seven hundred tomato plants, the winter kind. It was my dream; the only bad thing is that it has wind turbines. But I've got to tell you, you have to love [living in the *mas*], you have to be at ease doing this, otherwise you don't do it, because it is hard. I really like it. I rented it for eight years. I put a few solar panels, so I don't pay electricity, and water is cheap, because I use the water from the irrigation system, and food is never missing on the table. And I keep doing little jobs, a little bit of pruning over here, a few trees over there, just keeping going.

Although Iban complained about the wind turbines, the fact is that those turbines had made it possible for him to become a *masover*. The *mas* was owned by Fidel, the antinuclear activist and local lawyer (chapter 3), who used the *mas* exclusively for recreation purposes, as a purely leisure-oriented *locus amoenus*. In addition to the *mas* and a patch of forest, the parcel of land included three hectares of almond trees and vines, but Fidel never farmed those directly, hiring wage labor for that task instead. Fidel became deeply involved in the opposition to the wind farms, and once they were built, he stopped going to the *mas*: "I can't go. For me that was a place of relaxation and fun, and now when I see the turbines I get anxious, they've destroyed that place."

Iban had worked as a temporary wage laborer in a vineyard next to the *mas*, and he noticed that no one was using it—"the pine trees were already getting inside the *mas*." He got in touch with Fidel and they agreed on the rental: two hundred fifty euros a month, with an option that would allow Iban to buy the *mas* once the contract expired.

When I went to see the *mas* on the weekend, I was impressed. Iban had been living in the *mas* for just five months, and he had already put together a coop for a dozen hens and another for three dozen rabbits, had built planters filled with blooming flowers, and planted a huge vegetable garden, all irrigated. His aunt taught him how to slaughter the rabbits, and we ate one for lunch, cooked in an old wood-fueled stove that belonged to some relatives.

The planters and the chicken coop were built with drystone, as were the stairs that connected the *mas* with the garden, where I spotted a bench Iban had made from the pines that he had cut to clear the house entrance. "I did it with a few trunks and branches. It makes for a good seat. You'll see, sit down on it. In the afternoon it is ideal, because you get the shade from the tree. So when I am in the garden and I want to rest for a moment, I come up here and I have a smoke, *tranquil*."

He continued: "You know, I'm thirty-three, and in these five months I've learnt more about myself than in the eight years that I lived in Reus. Mind you, it's hard, there's a lot of time to be alone and think; you must like it. I was working for this important construction company: upward of 2,000 [Euros] a month, mobile phone, car, and then it all ended. Just like that. And now it's over, and it's over for me but also for those who had been working there for thirty years."

Then, he pointed at the garden: "The garden is the most important: food on the table. I always have potatoes, tomatoes, zucchini, eggplant, peppers, lettuce, onion, garlic, eight or nine eggs every day, one or two rabbits a week. I can't eat it all, it is to share with my parents and brother, it's a family operation, and it only works if we are together. And then I planted all these tomato plants with the idea of selling them, and maybe in the future I'll do an organic basket for city people."

In the following weeks and years, I visited Iban in the *mas* on numerous occasions. I was always offered tasty meals, and I always left with a bag filled with vegetables from the garden, ever in full production. Iban's mother and her brother, from a landless family, spent most mornings in the *mas*, cleaning and fixing things. His father and brother, also temporary agrarian workers, often had lunch in the *mas*. On each new visit, I was shown a new improvement. Once, Iban had borrowed a metal detector and collected an astounding quantity of explosives and other war material from the forest, repurposing it as decoration on the planters and the house entry. Another time he had built what he called a "chill-out area" by the *mas*: he converted two industrial wood spools into picnic tables, enclosed them with a sackcloth from the olive harvest to act as a mosquito net, and decorated the adjacent wall with old farming tools. There were also setbacks: the police came and seized the war material, Iban's dogs ate a bunch of the rabbits, and the first tomato harvest was lost due to an early freeze.

One day Iban and I were having coffee with a couple of friends of his in the chill-out area. There was a soft breeze, the conversation was engaging, and Iban said: "This is quality of life."

I agreed, adding something like "You bet, you've built a little paradise here." My comment offended Iban:

This is quality of life, but not luxury. In the *mas* and off the land you don't live, you bad-live. Write this down: the people who built this *mas* lived here because they wanted to, I live here because I need to (*ells hi vivien perquè els sortia dels collons, jo hi visc per collons*). This is the difference. I was tired of little jobs that took me nowhere; I don't want to suffer. I am tired of going to the bar. You only enjoy the *mas* if you know how to suffer. When I come here, I say to myself, "I am going to Korea."

Iban was tired of suffering, but stressed that one can only enjoy the *mas* through suffering. The suffering of the *mas* is different from the suffering of being jobless. He is making his place. The *mas* has become a *locus amoenus* and a place of resistance, both as a resource base and as a platform from which to give direction to his precarious condition. Iban's attempts to make money from the *mas* have thus far been unsuccessful; all his income comes from temporary construction and farming jobs, and he is exploring the possibility of working in the refueling operation at the nuclear plant. Nonetheless, the *mas* is central to his livelihood.

Every time Iban showed me an improvement in the *mas* he offered the same comment.

I heard it so many times that it almost felt like a joke: "It looks good, doesn't it? And the most important thing: *ni un duro* (it didn't cost a nickel)." There was one notable exception: the additional solar panels and high-storage capacity batteries Iban added to the *mas*.

He values the autonomy, the capacity that energy self-provision gives him to manage his own time: "These panels and batteries are awesome. Now, I go and look and see that I've made 9,500 watts and used 9,100. 'Ok, I have 400,' so I know that I can use the TV for about ten hours."

One of the last times we spoke he was considering installing a small wind turbine as a backup for cloudy days. With regard to the turbines surrounding the *mas*, he was straightforward: "I don't care what people say: this is the future. They bother me, true, but they are the future. They keep going and going, and they are clean. I mean, I wish I could get the power from them."

On my last visit to Iban's *mas*, I noticed an old document framed nicely on the wall. It was a police report, denouncing Iban's grandfather for selling eight sacks of barley on the black market. The document was dated in 1951, the same year that Joaquim, Andreu's father, was almost killed by the police, as we saw in chapter 1, for illegally selling wood and then helping a person who was suspected of being a member of the *maquis*, the anti-Francoist resistance guerrilla.

"He was a great man, my grandfather. He spent five years in a concentration camp after the war, and he saved the lives of two people," said Iban.

"Do you mean during the war?" I asked.

"No, afterward," he replied, keeping his eyes on me, a subtle smile across his face, not another word from his mouth. They were *maquis*. Iban took a paper from a shelf. It was a certificate, recently issued, from the Catalan government, recognizing his grandfather as a victim of the fascist repression.

"Recognition is very important."

Iban keeps resisting.

Notes

1. The *aval* was a letter stating that the republican soldier was not politically dangerous. It had to be sent by a person who had a proven record of loyalty to the new regime.
2. See Gidwani (1992, 2008, 2012).
3. On how colonial visions of *terra nullius* are used in contemporary developmental programs see Whitehead (2010) and Makki and Geisler (2011).
4. Gidwani (1992: 39). In home soil, British political economists used similar moral injunctions to denounce the "sloth" of small producers (Perelman 2000).
5. Douglas (2002: 197).
6. Gidwani (2012: 275–277).
7. Ferry and Limbert (2008: 13).
8. On the modern European trope of nature as bountiful yet wasteful, see Williams (1980).
9. Gidwani (2012: 285).
10. Bonefeld (2011: 380).
11. Nichols (2015: 27).
12. Wright (2006: 104).
13. Gidwani (2012: 285).
14. See Scoones (2015) for a recent overview of the literature on livelihood strategies.
15. For a comparable classification applied to a northern Valencian county, see Moragues-Faus (2014).
16. See Bourdieu (2008).
17. Data on gross family revenue (usually called BFR), last edited October 15, 2017, last accessed October 3, 2017, http://www.idescat.cat.
18. Since the advent of the crisis, agrarian land prices have gone significantly down in most of Spain while the cultivated area has also been reduced, suggesting a process of peasant deactivation. However, Catalonia represents an exception to this trend: cultivated area has remained steady and prices have modestly gone up (all without a sizable increment of corporate farming), thus suggesting that deactivation coexists with repeasantization (Soler and Fernández 2015: 22–131).

19. On drystone architectural elements in Terra Alta, see Rebés (2003).

20. De Certeau (1984: xv).

21. Lévi-Strauss (1966: 17).

22. Gramsci (1971, especially 323–377); Wainwright (2010).

23. Overlap of the terms referring to "outside" and to "fields" or "country" is widespread in Indo-European languages. According to Benveniste this overlap conjures up the image of an ancient relationship: "the uncultivated ground, the waste land, as opposed to the inhabited area" (2016: 255–256). This offers further evidence that in Southern Catalonia "going/working outside" (i.e. farming) is understood as a continuous activity of socializing nature, of making the *ager* ("field, wilderness") cultural—*agri-culture*.

24. For some recent debates, see Rosen (2012), McCrudden (2013), and Waldron et al. (2015). See Lomnitz (2012), Nader (2013), and Narotzky (2016a) for three rare anthropological contributions to this issue.

25. Scott (2013).

26. Riley (2013: 105).

27. Bloch (1986: 220).

28. Cited in Bonefeld and Psychopedis (2005: 4); see also Gilabert (2017).

29. Marx and Engels (1978: 133-34; original emphasis).

30. Bonefeld (2001).

31. It is unsurprising that its use is especially widespread in social contexts where there is a perceived danger of falling into slavery. See Graeber (2011), Bromell (2013) and Scott (2013).

32. Bonefeld and Psychopedis (2005: 4–5).

33. Gudeman (1986: 157).

34. Williams (1977).

35. Bensaïd (2001: 106).

36. Bourdieu (1988: 87); see also Lordon (2013: especially 223–244).

37. Smith (2016: 228).

38. Thompson (1971).

39. Bromell (2013: 292).

40. Antentas (2015: 142).

41. Krauss (2010: 206).

42. See http://doterraalta.com/wp-content/uploads/2017/07/MANIFEST-TA-CONTRA -MASSIFICACIO-EOLICA.pdf.

Bibliography

Abramsky, Kolya, ed. 2010. *Sparking a Worldwide Energy Revolution: Social Struggles in the Transition to a Post-Petrol World*. London: AK Press.

Abramsky, Kolya and Massimo De Angelis. 2009. "Introduction: Energy Crisis (Among Others) is in the Air." *The Commoner* 13:1–14.

ACA (Asociación de Ciencias Ambientales). 2016. *Pobreza, Vulnerabilidad y Desigualdad Energética: Nuevos Enfoques de Análisis*. Madrid: ACA.

Acheson, James N. and Anne W. Acheson. 2016. "Offshore Wind Power Development in Maine: A Rational Choice Perspective." *Economic Anthropology* 3(1):145–160.

Adams, Richard N. 1975. *Energy and Structure: A Theory of Social Power*. Austin, TX: University of Texas Press.

AEE (Asociación Empresarial Eólica). 2016. *Eólica '16*. Madrid: Asociación Empresarial Eólica.

Aguilera Klink, Federico. 2003. "Gestión Autoritaria versus Gestión Democrática del Agua." *Archipiélago* 57:34–42.

Ajuntament de la Fatarella. 2005. *L'Agenda 21 de la Fatarella: Document de Síntesi del Procés*. Calaceit (Spain): Gràfiques del Matarranya.

——— 1994. *Diagnòstic i Proposes per al Desenvolupament Local del Municipi de la Fatarella*. Calaceit (Spain): Gràfiques del Matarranya.

Alanne, Kari and Arto Saari. 2006. "Distributed Energy Generation and Sustainable Development." *Renewable and Sustainable Energy Reviews* 10(6):539–558.

Alayo, Joan C. 2011. *L'electricitat a Catalunya, de 1876 a 1936*. Lleida (Spain): Pagès.

Alfama, Eva and Neus Miró (eds.). 2005. *Dones en Moviment: Una Anàlisi de Gènere de la Lluita en Defensa de l'Ebre*. Valls (Spain): Cossetània.

Alfama, Eva, Àlex Casademunt, Gerard Coll-Planas, Helena Cruz and Marc Martí. 2007. *Per una Nova Cultura del Territori? Mobilitzacions i Conflictes Territorials*. Barcelona: Icaria.

Almenar, Ricardo and Emèrit Bono. 2005. *Trasvase del Ebro y Comunidad Valenciana: Artículos, Conferencias e Informes para la Sostenibilidad Hídrica de un País Mediterráneo (2000–2004)*. Valencia (Spain): Tirant lo Blanch.

Alonso Biarge, José M. (coord). 1978. *Estudio de Alternativas de Aprovechamiento del Río Ebro y de Desarrollo Socioeconómico y Protección Ecológica de su Valle*. Tarragona (Spain): Diputación Provincial de Tarragona.

Altvater, Elmar. 1993. *The Future of the Market: An Essay on the Regulation of Money and Nature after the Collapse of "Actually Existing Socialism."* London: Verso.

Anderson, Kevin B. 2010. *Marx at the Margins: On Nationalism, Ethnicity, and Non-Western societies*. Chicago: University of Chicago Press.

Antentas, Josep M. 2015. "Spain: The *Indignados* Rebellion of 2011 in Perspective." *Labor History* 56(2):136–160.

Appadurai, Arjun. 1986. "Introduction: Commodities and the Politics of Value." In *The Social Life of Things: Commodities in Cultural Perspective*, edited by Arjun Appadurai, 3–63. Cambridge: Cambridge University Press.

Argenti, Nicolas and Daniel M. Knight. 2015. "Sun, Wind, and the Rebirth of Extractive Economies: Renewable Energy Investment and Metanarratives of Crisis in Greece." *Journal of the Royal Anthropological Institute* 21(4):781–802.

Ariza-Montobbio, Pere. 2013. *Large-Scale Renewable Energy? A Transdisciplinary View on Conflicts and Trade-offs in the Implementation of Renewable Energy.* Unpublished PhD dissertation, Universitat Autònoma de Barcelona, 217 pages.

Arrojo, Pedro. 2006. *El Reto Ético de la Nueva Cultura del Agua: Funciones, Valores y Derechos en Juego.* Barcelona: Paidós.

Asamblea de Cercedilla. 1979. "Por qué una Federación?" *Cuadernos del Ruedo Ibérico* 63–66:302.

Associació de Joves de la Fatarella. 1990. *Aniversari de l'Associació de Joves Amics de la Cultura de la Fatarella.* Calaceit (Spain): Gràfiques del Matarranya.

Audí Ferrer, Pere. 2010. *Cooperativistes, Anarquistes i Capellans al Priorat (1910–1923).* Torroja del Priorat (Spain): Arxiu Comarcal del Priorat.

Bailey, Anne M. 1990. "Rural Revolution in Catalonia during the Civil War: The Meaning of Multiple Sovereignty." *Critique of Anthropology* 10(2/3):179–196.

Ballard, Chris and Glenn Banks. 2003. "Resource Wars: The Anthropology of Mining." *Annual Review of Anthropology* 32:287–313.

Banaji, Jairus. 2016. "Merchant Capitalism, Peasant Households and Industrial Accumulation: Integration of a Model." *Journal of Agrarian Change* 16(3):410–431.

Barcelona, Ricardo G. 2012. "Wind Power Deployment—Why Spain Succeeded." *Renewable Energy Law and Policy* 2:146–149.

Barthes, Roland. 2015 [1957]. *Mythologies.* Paris: Seuil.

Bayerri, Josep. 1985. *Les Terres de l'Ebre, encara un Futur...?* Tortosa (Spain): Dertosa.

Benelbas, León, Xavier Garcia, and Joan Tudela. 1977. *La Unió de Pagesos, el Sindicat del Camp.* Barcelona: 7x7 Edicions.

Benjamin, Walter. 1968a [1942]. "Theses on the Philosophy of History." In *Iluminations: Essays and Reflections*, edited by Hannah Arendt, 253–264. New York: Schocken Books.

——— 1968b [1936]. "The work of Art in the Age of Mechanical Reproduction." In *Iluminations: Essays and Reflections*, edited by Hannah Arendt, 217–251. New York: Schocken Books.

Bensaïd, Daniel. 2001. *Les Irreductibles: Théorèmes de la Résistance à l'Air du Temps.* Paris: Textuel.

Benveniste, Émile. 2016 [1969]. *Dictionary of Indo-European Concepts and Society.* Chicago: Hau Books.

Bernstein, Henry. 2010. *Class Dynamics of Agrarian Change.* Black Point, NS: Fernwood Publishing.

Bloch, Ernst. 1991 [1962]. *Heritage of Our Times.* Cambridge: Polity Press.

——— 1986. *Natural Law and Human Dignity.* Cambridge, MA: MIT Press.

Bloch, Maurice and Jonathan Parry. 1989. "Introduction: Money and the Morality of Exchange." In *Money and the Morality of Exchange*, edited by Maurice Bloch and Jonathan Parry, 1–32. Cambridge: Cambridge University Press.

Blowers, Andrew and Pieter Leroy. 1994. "Power, Politics, and Environmental Inequality: A Theoretical and Empirical Analysis of the Process of 'Peripheralisation'." *Environmental Politics* 3(2):197–228.

Bohannan, Paul. 1959. "The Impact of Money on an African Subsistence Economy." *The Journal of Economic History* 19(4):491–503.

Bonefeld, Werner. 2011. "Primitive Accumulation and Capitalist Accumulation: Notes on Social Constitution and Expropriation." *Science & Society* 75(3):379–399.

——— 2001. "Social Form, Critique and Human Dignity." *Zeitschrift für Kritische Theorie* 13:97–112.

Bonefeld, Werner and Kosmas Psychopedis, eds. 2005. *Human Dignity: Social Autonomy and the Critique of Capitalism*. Aldershot (UK): Ashgate.

Boquera Margalef, Montserrat. 2009. *Primer la Sang que l'Aigua: Els Pilars d'una Nova Identitat Ebrenca*. Benicarló (Spain): Onada Edicions.

Bourdieu, Pierre. 2008 [2002]. *The Bachelor's Ball: The Crisis of Peasant Society in Béarn*. Cambridge: Polity Press.

——— 1988. *Homo Academicus*. Palo Alto, CA: Stanford University Press.

——— 1977. *Outline of a Theory of Practice*. Cambridge: Cambridge University Press.

Boyer, Dominic. 2014. "Energopower: An Introduction." *Anthropological Quarterly* 87(2):309–334.

——— 2011. "Energopolitics and the Anthropology of Energy." *Anthropology News* 52(5):5–7.

Brennan, Timothy. 2017. "On the Image of the Country and the City." *Antipode* 49(S1):34–51.

Brenner, Neil and Stuart Elden. 2009. "State, Space, World: Lefebvre and the Survival of Capitalism." In *State, Space, World: Selected Essays*, edited by Neil Brenner and Stuart Elden, 1–48. Minneapolis, MN: University of Minnesota Press.

Bretón, Víctor. 2000. *Tierra, Estado y Capitalismo: La Transformación Agraria del Occidente Catalán (1940–1990)*. Lleida (Spain): Milenio.

Bridge, Gavin, Stefan Bouzarovski, Michael Bradshaw, and Nick Eyre. 2013. "Geographies of Energy Transition." *Energy Policy* 53:331–340.

Bromell, Nick. 2013. "Democratic Indignation: Black American Thought and the Politics of Dignity." *Political Theory* 41(2):285–311.

Browne, Katherine E. 2009. "Economics and Morality: Introduction." In *Economics and Morality: Anthropological Approaches*, edited by Katherine E. Browne and B. Lynne Milgram, 1–40. Lanham, MD: AltaMira.

Bunker, Stephen G. 1984. "Modes of Extraction, Unequal Exchange, and the Progressive Underdevelopment of an Extreme Periphery: The Brazilian Amazon, 1600–1980." *American Journal of Sociology* 89(5):1017–1064.

Burawoy, Michael. 2013. "Marxism after Polanyi." In *Marxisms in the 21st Century*, edited by Michelle Williams and Vishwas Satgar, 34–52. Johannesburg: Wits University Press.

Caffentzis, George. 2013. *In Letters of Blood and Fire: Work, Machines, and the Crisis of Capitalism*. Oakland, CA: PM Press.

——— 1992 [1980]. "The Work/Energy Crisis and the Aplocalypse." In *Midnight Oil: Work, Energy, War, 1972–1992*, edited by Midnight Notes Collective, 215–271. Brooklyn, NY: Autonomedia.

Canosa, Francesc. 2014. *Capitans d'Indústria Explicats pels Seus Fills*. Barcelona: Mobil Books.

Capel, Horacio (ed). 1994. *Las Tres Chimeneas: Implantación Industrial, Cambio Tecnológico y Transformación de un Espacio Urbano Barcelonés*. 3 Volumes. Barcelona: FECSA.

Capel, Horacio and Ignacio Muro. 1994. "La Central de Mata y el Nuevo Papel de la Energía Térmica, 1951–1974." In *Las Tres Chimeneas*, Volume 3, edited by Horacio Capel, 13–75. Barcelona: FECSA.

Capel, Horacio and Luis Urteaga. 1994. "El Triunfo de la Hidroelectricidad y la Expansion de 'La Canadiense'." In *Las Tres Chimeneas*, Volume 2, edited by Horacio Capel, 13–81. Barcelona: FECSA.

Carpintero, Óscar (dir). 2015. *El Metabolismo Económico Regional Español*. Madrid: FUHEM Ecosocial.

Carreras, Albert and Xavier Tafunell. 2010. *Historia Económica de la España Contemporánea (1789–2009)*. Barcelona: Crítica.

Casanova, Julián and Carlos Gil Andrés. 2014 [2009]. *Twentieth Century Spain: A History*. Cambridge: Cambridge University Press.

Castell, Edmon et al. 1999. *La Batalla de l'Ebre: Història, Paisatge, Patrimoni*. Barcelona: Pòrtic.

Castell, Edmon and Oriol Nel·lo. 2003. "El Parc Eòlic de les Serres de Pàndols i Cavalls: Energia, Valors Ambienals i Memoria Històrica." In *Aquí no! Els Conflictes Territorials a Catalunya*, edited by Oriol Nel·lo, 69–93. Barcelona: Empúries.

Charnock, Greig, Thomas Purcell and Ramon Ribera-Fumaz. 2011. "¡Indígnate!: The 2011 Popular Protests and the Limits to Democracy in Spain." *Capital & Class* 36(1):3–11.

Christopherson, Susan and Ned Rightor. 2012. "How Shale Gas Extraction Affects Drilling Localities: Lessons for Regional and City Policy Makers." *Journal of Town & City Management* 2(4):350–368.

Clar Moliner, Ernesto and Javier Silvestre Rodríguez. 2008. "Impactos Demográficos." In *Gestión y Usos del Agua en la Cuenca del Ebro en el Siglo XX*, edited by Vicente Pinilla Navarro, 657–673. Zaragoza (Spain): Prensas Universitarias de Zaragoza.

Clark, Brett and John B. Foster. 2012. "Guano: The Global Metabolic Rift and the Fertilizer Trade." In *Ecology and Power: Struggles over Land and Material Resources in the Past, Present and Future*, edited by Alf Hornborg et al, 68–82. London: Routledge.

Cleaver, Harry. 1992. "The Inversion of Class Perspective in Marxian Theory: From Valorization to Self-Valorization." In *Essays on Open Marxism*, edited by Werner Bonefeld et al, 106–144. London: Pluto Press.

Collins, Jane. 2017. *The Politics of Value: Three Movements to Change How We Think about the Economy*. Chicago: University of Chicago Press.

——— 2011. "What Difference Does Financial Expansion Make? A Response to Gavin Smith's Paper 'Selective Hegemony and Beyond'." *Identities* 18(1):39–46.

Contreras, Jesús. 1991a. "Los Grupos Domesticos: Estrategias de Producción y de Reproducción." In *Antropología de los Pueblos de España*, edited by Joan Prat et al., 343–380. Madrid: Taurus.

——— 1991b. "Estratificación Social y Relaciones de Poder." In *Antropología de los Pueblos de España*, edited by Joan Prat et al., 499–519. Madrid: Taurus.

Coronil, Fernando. 1997. *The Magical State: Nature, Money and Modernity in Venezuela*. Chicago: University of Chicago Press.

Costa, Joaquín. 1901. *Oligarquía y Caciquismo como la Forma Actual del Gobierno en España*. Madrid: Real Academia de la Historia.

Cottrell, Frederick W. 1955. *Energy and Society: The Relation between Energy, Social Change, and Economic Development*. New York: McGraw-Hill.

Cronon, William. 1991. *Nature's Metropolis: Chicago and the Great West*. New York: Norton & Company.

Cucó, Josepa. 1982. *La Tierra como Motivo: Jornaleros y Propietarios en Dos Pueblos Valencianos*. Valencia: Alfons el Magnànim.

Cuerdo Mir, Miguel. 1999. "Evaluación de los Planes Energéticos Nacionales en España (1975–1998)." *Revista de Historia Industrial* 15:161–178.

De Angelis, Massimo. 2007. *The Beginning of History: Value Struggles and Global Capital.* London: Pluto.

De Certeau, Michel. 1984 [1980]. *The Practice of Everyday Life.* Berkeley, CA: University of California Press.

De la Cadena, Marisol. 2015. *Earth Beings: Ecologies of Practice across Andean Worlds.* Durham, NC: Duke University Press.

De Prada, Carlos. 2003. "La 'Racionalidad Oculta' del Plan Hidrológico Nacional." *Archipiélago* 57:58–68.

Debeir, Jean Claude, Jean-Paul Deléage, and Daniel Hemery. 1991 [1986]. *In the Servitude of Power: Energy and Civilization through the Ages.* London: Zed Books.

Del Moral Ituarte, Leandro. 2003. "Planificación Hidrológica, Mercado y Territorio." *Archipiélago* 57:9–16.

Descola, Philippe. 2013. *Beyond Nature and Culture.* Chicago: University of Chicago Press.

Domènech, Antoni. 2004. *El Eclipse de la Modernidad: Una Revisión Republicana de la Tradición Socialista.* Barcelona: Crítica.

Douglas, Mary. 2002 [1966]. *Purity and Danger.* London: Routledge.

Dracklé, Dorle and Werner Krauss. 2011. "Ethnographies of Wind and Power: Energy and Energopolitics." *Anthropology News* 52(5):5–10.

Dupuy, Jean-Pierre. 1999. *El Pánico.* Barcelona: Gedisa.

Edelman, Marc. 2005. "Bringing the Moral Economy back in … to the Study of 21st-Century Transnational Peasant Movements." *American Anthropologist* 107(3):331–345.

Elson, Diane. 1979. "The Value Theory of Labour." In *The Representation of Labour in Capitalism,* edited by Diane Elson, 115–180. Atlantic Highlands, NJ: Humanities Press.

Epstein, Arnold L. 1958. *Politics in an Urban African Community.* Manchester (UK): Manchester University Press.

Escobar, Arturo. 2006. "Difference and Conflict in the Struggle over Natural Resources: A Political Ecology Framework." *Development* 49(3):6–13.

Estevan, Antonio. 2003. "El Plan Hidrológico Nacional: Destapando la Olla." *Archipiélago* 57:43–57.

Etxezarreta, Miren (coord). 2006. *La Agricultura Española en la Era de la Globalización.* Madrid: Ministerior de Agricultura, Pesca y Alimentación.

———1991. "La Modernization de l'Agriculture Espagnole et le Développement Rural." *Économie Rurale* 202/203:44–45.

Etxezarreta, Miren, Josefina Cruz, Mario García Morilla, and Lourdes Viladomiu. 1995. *La Agricultura Familiar, ante las Nuevas Políticas Agrarias.* Madrid: Ministerio de Agricultura, Pesca y Alimentación.

European Commission. 2011. "Energy 2020—A Strategy for Competitive, Sustainable and Secure Energy." *Renewable Energy Law and Policy* 2(1):50–74.

Fabra Portela, Natalia and Jorge Fabra Utray. 2012. "El Déficit Tarifario en el Sector Eléctrico Español." *Papeles de Economía Española* 134:88–100.

Fabra Utray, Jorge. 2014. "Los Infundados Fundamentos de la Regulación Eléctrica Vigente." In *Alta Tensión,* edited by JV Barcia Magaz and Cote Romero, 83–96. Barcelona: Icaria.

Fassin, Didier. 2009. "Les Économies Morales Revisitées." *Annales: Histoire, Sciences Sociales* 64(6):1237–1266.

Ferguson, James. 2005. "Seeing Like an Oil Company: Space, Security, and Global Capital in Neoliberal Africa." *American Anthropologist* 107(3):377–382.

Fernández, Joaquín. 1999. *El Ecologismo Español*. Madrid: Alianza.

Fernández Durán, Ramón. 2006. *El Tsunami Urbanizador Español y Mundial*. Barcelona: Virus.

Ferrús, Jordi. 2004. *Transición Social en Catalunya Nova: El Impacto de la Instalación de una Central Nuclear en la Economia y el Poder Local (1970–1990)*. Unpublished PhD dissertation. Department of Social Anthropology, Philosophy and Social Work, Universitat Rovira i Virgili (Tarragona, Spain), 1042 pages.

—— 1985. *La Casa Pagesa Asconenca, 1940–1970: Descripció Etnogràfica, Anàlisi Antropològica*. Flix (Spain): Centre d'Estudis de la Ribera d'Ebre.

Ferry, Elisabeth E. 2002. "Inalienable Commodities: The Production and Circulation of Silver and Patrimony in a Mexican Mining Cooperative." *Cultural Anthropology* 17(3):331–358.

Ferry Elizabeth E. and Mamdana E. Limbert. 2008. "Introduction." In *Timely Assets: The Politics of Resources and Their Temporalities*, edited by Elizabeth E. Ferry and Mamdana E. Limbert, 3–24. Santa Fe (NM): SAR Press.

Firat, Bilge. 2014. "Crisis, Power and Policymaking in the New Europe: Why Should Anthropologists Care?" *Anthropological Journal of European Cultures* 23(1):1–20.

Fischer-Kowalski, Marina. 1998. "Society's Metabolism: The Intellectual History of Materials Flow Analysis, Part I, 1860–1970." *Journal of Industrial Ecology* 2(1):61–136.

Flyvbjerg, Bent. 2005. "Machiavellian Megaprojects." *Antipode* 37(1):18–22.

Foster, John B. 2000. *Marx's Ecology: Materialism and Nature*. New York: New York University Press.

Franquesa, Jaume. 2016. "Dignity and Indignation: Bridging Morality and Political Economy in Contemporary Spain." *Dialectical Anthropology* 40(2):69–86.

—— 2014. "Consolidating Power, Controlling the Future: Debt and Crisis in the Spanish Electrical Sector." *Focaalblog* (July 17).

—— 2013a. *Urbanismo Neoliberal, Negocio Inmobiliario y Vida Vecinal: El Caso de Palma*. Barcelona: Icaria.

—— 2013b. "On Keeping and Selling: The Political Economy of Heritage Making in Contemporary Spain." *Current Anthropology* 54(3):346–369.

Friedman, Jonathan. 1974. "The Place of Fetishism and the Problem of Materialist Interpretations." *Critique of Anthropology* 1(1):26–62.

Frigolé, Joan. 1995. *Un Etnólogo en el Teatro: Ensayo Antropológico sobre Federico García Lorca*. Barcelona: Muchnik.

—— 1991. "'Ser Cacique' y 'Ser Hombre' o la Negación de las Relaciones de Patronazgo en un Pueblo de la Vega Alta del Segura." In *Antropología de los Pueblos de España*, edited by Joan Prat et al., 556–573. Madrid: Taurus.

Fucho, Felip. 1998. *El llibre de la Caça: Els Sistemes de Caça a la Fatarella, Terra Alta*. Barcelona: Columna.

Gallego, Cristóbal J. and Marta Victoria. 2012. *Entiende el Mercado Eléctrico*. Madrid: Observatorio Crítico de la Energía.

Gallego, Ferran. 2008. *El Mito de la Transición*. Barcelona: Crítica.

Garcés, Joan. 2012. *Soberanos e Intervenidos: Americanos y Españoles, Estrategias Globales*. Fourth edition. Madrid: Siglo XXI.

Garcia, Xavier. 2013. *Carmel d'Ascó: Una Visió Biogràfica*. Flix (Spain): CERE.

———— 2008. *La Primera Dècada de Lluita Antinuclear a Catalunya (1970–1980)*. Torroja del Priorat (Spain): Arxiu Comarcal del Priorat.

———— 2003. *Catalunya es Revolta*. Barcelona: Angle Editorial.

———— 1997. *Catalunya Tambe Té Sud*. Barcelona: Flor del Vent.

Garcia, Xavier, Jaume Rexach, and Santiago Vilanova. 1979. *El Combat Ecologista a Catalunya*. Barcelona: Edicions 62.

Garí Ramos, Manuel. 2014. "Evolución del Empleo en las Energías Renovables en el Estado Español (2004–2012)." In *Alta Tensión*, edited by JV Barcia Magaz and Cote Romero, 195–210. Barcelona: Icaria.

Garrabou, Ramon, ed. 2006. *Història Agrària dels Països Catalans IV: Segles XIX–XX*. Barcelona: Fundació Catalana per a la Recerca.

Garrabou, Ramon and Enric Saguer. 2006. "Propietat, Tinença i Relacions de Distribució." In *Història Agrària dels Països Catalans IV: Segles XIX–XX*, edited by Ramon Garrabou, 353–431. Barcelona: Fundació Catalana per a la Recerca.

Gaviria, Mario, ed. 1977. *El Bajo Aragón Expoliado: Recursos Naturales y Autonomía Regional*. Zaragoza (Spain): DEIBA.

Geddes, Patrick. 1968 [1915]. *Cities in Evolution: An Introduction to the Town Planning Movement and the Study of Civics*. New York: Harper & Row.

Gellner, Ernest and John Waterbury (eds.). 1977. *Patrons and Clients in Mediterranean Societies*. London: Duckworth.

Gidwani, Vinay K. 2012. "Waste/Value." In *The Wiley-Blackwell Companion to Economic Geography*, edited by Trevor J. Barnes, Jamie Peck and Eric Shepard, 275–288. Maldon, MA: Wiley-Blackwell.

———— 2008. *Capital, Interrupted: Agrarian Development and the Politics of Work in India*. Minneapolis, MN: University of Minnesota Press.

———— 1992. "'Waste' and the Permanent Settlement in Bengal." *Economic and Political Weekly* 27(4):39–46.

Gilabert, Pablo. 2017. "Kantian Dignity and Marxian Socialism." *Kantian Review* 22(4):553–577.

Gill, Leslie and Sharryn Kasmir. 2016. "History, Space, Labor: On Unevenness as an Anthropological Concept." *Dialectical Anthropology* 40(2):87–102.

Gluckman, Max. 1963. "Rituals of Rebellion in Southeast Africa." In *Order and Rebellion in Tribal Africa*, 110–136. New York: Free Press.

Godelier, Maurice. 2001 [1984]. *The Mental and the Material*. London: Verso.

Gómez Mendoza, Antonio, Carles Sudrià, and Javier Pueyo. 2007. *Electra y el Estado: La Intervención Pública en la Industria Eléctrica bajo el Franquismo*. 2 Volumes. Madrid: Comisión Nacional de la Energía and Thomson-Civitas.

González, Erika, Kristina Sáez, and Pedro Ramiro. 2010. "Multinational Companies and the Energy Crisis in Latin America." In *Sparking a Worldwide Energy Revolution*, edited by Kolya Abramsky, 178–187. London: AK Press.

Graeber, David. 2011. *Debt: The First 5000 Years*. New York: Melville House.

———— 2001. *Toward an Anthropological Theory of Value: The False Coin of Our Own Dreams*. New York: Palgrave.

Gramsci, Antonio. 1971. *Selections from the Prison Notebooks*. New York: International Publishers.

———— 1957 [1926]. "The Southern Question." In *The Modern Prince & Other Writings*, 28–51. New York: International Publishers.

Grau Folch, Josep J. 1993. *La Terra Alta: Estructures Productives i Evolució Social*. Barcelona: Caixa Catalunya.

Gros, Alba. 2010. *Terra i Pedra: Agricultura Tradicional de Secà*. Fatarella: Fundació El Solà.

Gudeman, Stephen. 1986. *Economics as Culture: Models and Metaphors of Livelihood*. London: Routledge & Kegan Paul.

Guiu, Claire. 2008. *Naissance d'une Autre Catalogne*. Paris: Editions CTHS.

Hajer, Maarten A. 1995. *The Politics of Environmental Discourse: Ecological Modernization and the Policy Process*. Oxford: Clarendon Press.

Haraway, Donna and Cary Wolfe. 2016. *Manifestly Haraway*. Minneapolis, MN: University of Minnesota Press.

Harootunian, Harry. 2015. *Marx after Marx: History and Time in the Expansion of Capitalism*. New York: Columbia University Press.

Harvey, David. 2001 [1974]. "Population, Resources and the Ideology of Science." In *Spaces of Capital: Towards a Critical Geography*, 38–67. New York: Routledge.

——— 1999 [1982]. *The Limits to Capital*. London: Verso.

Hecht, Gabrielle. 1994. *The Radiance of France: Nuclear Power and National Identity after World War II*. Cambridge, MA: MIT Press.

Herzfeld, Michael. 1991. *A Place in History: Social and Monumental Time in a Cretan Town*. Princeton: Princeton University Press.

Hinkelbein, Oliver. 2010. "Fiebre del Oro en la Era del Cambio Climático: El Mar del Norte como Potencia Eólica Emergente (Alemania)." *Nimbus* 25:111–129.

Horkheimer, Max and Theodor W. Adorno. 2002 [1947]. *The Dialectic of Enlightenment: Philosophical Fragments*. Stanford, CA: Stanford University Press.

Hornborg, Alf. 2016. *Global Magic: Technologies of Appropriation from Ancient Rome to Wall Street*. Basingstoke, UK: Palgrave Macmillan.

——— 2013. "The Fossil Interlude: Euro-American Power and the Return of the Physiocrats." In *Cultures of Energy: Power, Practices, and Technologies*, edited by Sarah Strauss et al., 41–59. Walnut Creek, CA: Left Coast Press.

——— 2011. *Global Ecology and Unequal Exchange: Fetishism in a Zero-Sum World*. London: Routledge.

——— 2001. *The Power of the Machine: Global Inequalities of Economy, Technology and Environment*. Walnut Creek, CA: AltaMira.

Howard, Ebenezer. 1965 [1902]. *Garden Cities of To-morrow*. Cambridge, MA: MIT Press.

Howe, Cymene. 2014. "Anthropocenic Ecoauthority: The Winds of Oaxaca." *Anthropological Quarterly* 87(2):381–404.

Huber, Matthew T. 2014. *Lifeblood: Oil, Freedom, and the Forces of Capital*. Minneapolis, MN: University of Minnesota Press.

Hughes, David. 2017. *Energy without Conscience: Oil, Climate Change and Complicity*. Durham, NC: Duke University Press.

Hughes, Thomas P. 1983. *Networks of Power: Electrification in Western Society*. Baltimore: Johns Hopkins University Press.

Hvelplund, Frede. 2014. "Black or Green Wind Power." In *Wind Power for the World: International Reviews and Developments*, edited by Preben Maegaard et al, 79–90. Singapore: Pan Stanford Publishing.

Ingold, Tim. 2000. "Globes and Spheres: The Topology of Environmentalism." In *The Perception of the Environment: Essays on Livelihood, Dwelling and Skill*, 209–218. London: Routledge.

Isenhour, Cindy and Kuishuang Feng. 2016. "Decoupling and Displaced Emissions: On Swedish Consumers, Chinese Producers and Policy to Address the Climate Impact of Consumption." *Journal of Cleaner Production* 134:320–329.

Izquierdo Martín, Jesús and Pablo Sánchez León. 2010. "El Agricultor Moral: Instituciones, Capital Social y Racionalidad en la Agricultura Española Contemporánea." *Revista Española de Estudios Agrosociales y Pesqueros* 225:137–169.

Jessop, Bob. 2008. *State Power: A Strategic-Relational Approach*. Cambridge: Polity Press.

Jochimsen, Ulrich. 2013. "How the Electricity Feed-in Law Came to Be Passed by the German Parliament, Enabling Renewable Energies to Establish Their Position in the Market." In *Wind Power for the World: International Reviews and Developments*, edited by Preben Maegaard et al., 479–490. Singapore: Pan Stanford Publishing.

Jociles Rubio, M. Isabel. 1989. *La Casa en la Catalunya Nova*. Madrid: Ministerio de Cultura.

Kalb, Don, and Herman Tak. 2005. *Critical Junctions: Anthropology and History Beyond the Cultural Turn*. New York: Berghahn Books.

Kasmir, Sharryn. 1996. *The "Myth" of Modragón: Cooperatives, Politics, and Working-class in a Basque Town*. New York: SUNY Press.

Kopytoff, Igor. 1986. "The Cultural Biography of Things: Commoditization as Process." In *The Social Life of Things: Commodities in Cultural Perspective*, edited by Arjun Appadurai, 64–91. Cambridge: Cambridge University Press.

Koster, Auriane M. and John M. Anderies. 2013. "Institutional Factors that Determine Energy Transitions: A Comparative Case Study Approach." In *Renewable Energy Governance*, edited by E. Michalena and J.M. Hills, 33–61. London: Springer.

Krauss, Werner. 2010. "The 'Dingpolitik' of Wind Energy in Northern German Landscapes: An Ethnographic Case Study." *Landscape Research* 35(2):195–208.

Kropotkin, Petr A. 1985 [1901]. *Fields, Factories and Workshops, or Industry Combined with Agriculture and Brain Work with Manual Work*. London: Freedom Press.

Larrú, José María. 2009. "El 'Caso España': Un Repaso a la Ayuda Norteamericana Recibida por España en Perspectiva Actual y Comparada." *Estudios Económicos de Desarrollo Internacional* 9(1):91–126.

Latouche, Serge. 2009. *Farewell to Growth*. Cambridge: Polity Press.

Latour, Bruno. 2017. *Facing Gaia: Eight Lectures on the New Climactic Regime*. Cambridge: Polity Press.

Lefebvre, Henri. 2009 [1979]. "Comments on a New State Form." In *State, Space, World. Selected Essays*, edited by Stuart Elden and Neil Brenner, 124–137. Minneapolis, MN: University of Minnesota Press.

—— 2004 [1992]. *Rythmanalysis: Space, Time and Everyday Life*. London: Continuum.

—— 1991 [1974]. *The Production of Space*. Oxford: Blackwell.

—— 1978. *De l'État. Volume 4: Les Contradictions de l'État Modern (la Dialectique et/de l'État)*. Paris: PUF.

Lenin, Vladimir I. 1978 [1894]. "What 'the Friends of the People' Are and How They Fight the Social Democrats." Available at: https://www.marxists.org/archive/lenin/works/1894/friends/index.htm.

Lévi-Strauss, Claude. 1966 [1962]. *The Savage Mind*. Chicago: University of Chicago Press.

Lewis, C. S. 2001 [1944]. *The Abolition of Man*. New York: HarperCollins.

Li, Tania. 2010. "To Make Live or Let Die? Rural Dispossession and the Protection of Surplus Populations." *Antipode* 41(S1):66–93.

Lleonart, Pere, Àlvar Garola, and Gemma Vélez. 2006. *Avaluació Socioeconòmica dels Parcs Eòlics Associats a AERTA a la Terra Alta*. Unpublished report, sponsored by AERTA, 79 pages.

Logan, John R. 1985. "Democracy from Above: Limits to Change in Southern Europe." In *Semiperipheral Development: The Politics of Southern Europe in the Twentieth Century*, edited by Giovanni Arrighi, 149–178. London: Sage.

Lomnitz, Claudio. "Dignidad." *La Jornada*, 11 January 2012.

López, Isidro and Emmanuel Rodríguez. 2011. "The Spanish Model." *New Left Review* 69:5–29.

—— 2010. *Fin de Ciclo: Financiarización, Territorio y Sociedad de Propietarios en la Onda Larga del Capitalismo Hispano (1959–2010)*. Madrid: Traficantes de sueños.

López Linaje, Javier. 1979. "Opciones Energéticas y Condicionantes Sociales: Concentración, Dependencia y Despilfarro frente a la Dispersión, Autonomía y Aprovechamiento Integral de la Energía Renovable." *Cuadernos del Ruedo Ibérico* 63–66:93–126.

Lordon, Frédéric. 2002. *La Politique du Capital*. Paris: Odile Jacob.

—— 2013. *La Société des Affects: Pour un Structuralisme des Passions*. Paris: Seuil.

Love, Thomas and Cindy Isenhour. 2016. "Energy and Economy: Recognizing High-Energy Modernity as a Historical Period." *Economic Anthropology* 3(1):6–16.

Lovins, Amory B. 1977. *Soft Energy Paths: Toward a Durable Peace*. Cambridge, MA: Ballinger.

Lovins, Amory, L. Hunter Lovins, and Paul Hawken. 1999. "A Road Map for Natural Capitalism." *Harvard Business Review* 1999 (May–June):145–158.

Lovins, Hunter. 1988. "A World for Generations to Come." In *Broken Circle: A Search for Wisdom in the Nuclear Age*, edited by Stephen Most and Lynn Grasberg. Palo Alto, CA: Consulting Psychologists Press.

Lowy, Michael. 2005. *Fire Alarm: Reading Walter Benjamin's "On the Concept of History."* London: Verso.

Maegaard, Preben, Anna Krenz, and Wolfgang Palz. 2013. *Wind Power for the World: The Rise of Modern Wind Energy*. Singapore: Pan Stanford Publishing.

Majoral, Roser. 2006. "De la Guerra Civil a la Unió Europea." In *Història Agrària dels Països Catalans IV: Segles XIX–XX*, edited by Ramon Garrabou, 605–651. Barcelona: Fundació Catalana per a la Recerca.

Makki, Fouad and Charles Geisler. 2011. "Development by Dispossession: Land Grabbing as New Enclosures in Contemporary Ethiopia." Paper presented at the International Conference on Global Land Grabbing, April 6–8, 2011, 21 pages.

Malm, Andreas. 2016. *Fossil Capital: The Rise of Steam Power and the Roots of Global Warming*. London: Verso.

Margalef, Joaquim and Joan Tasias. 1985. *El Priorat: Anàlisi d'una Crisi Productiva*. Barcelona: Caixa d'Estalvis de Catalunya.

Martín Arancibia, Salvador. 1979. "Energía y Política: Los Engaños del Plan Energético Nacional. *Cuadernos del Ruedo Ibérico* 63–66:269–306.

Martínez, Ladislao. 2014. "Productores de Nuestra Propia Energía: Plantas Colectivas y Autoconsumo." In *Alta Tensión*, edited by JV Barcia Magaz and Cote Romero, 239–248. Barcelona: Icaria.

Martínez Alier, Joan. 2009. "Social Metabolism, Ecological Distribution Conflicts, and Languages of Valuation." *Capitalism Nature Socialism* 20(1):58–87.

—— 2003. *The Environmentalism of the Poor: A Study of Ecological Conflicts and Valuation*. Cheltenham (UK): Edward Elgar.

—— 2002. *The Environmentalism of the Poor*. Paper prepared for the conference on "The Political Economy of Sustainable Development: Environmental Conflict, Participation and Movements." Geneva: UNRISD, 56 pages.

—— 1987. *Ecological Economics: Energy, Environment and Society*. Oxford: Blackwell.

Martínez Gil, Javier. 1997. *La Nueva Cultura del Agua en España*. Bilbao (Spain): Bakeaz.

Martínez-Val, José M. 2001. *Un Empeño Industrial que Cambió España, 1850–2000: Siglo y Medio de Ingeniería Industrial*. Madrid: Síntesis.

Martínez Veiga, Ubaldo. 1991. "Organización y Percepción del Espacio." In *Antropología de los Pueblos de España*, edited by Joan Prat et al., 195–255. Madrid: Taurus.

Marx, Karl. 1976 [1867]. *Capital*. Volume I. London: Penguin Books.

—— 1963 [1869]. *The 18th Brumaire of Louis Bonaparte*. New York: International Publishers.

Marx, Karl and Friedrich Engels. 1986 [1848]. *The Communist Manifesto*. Toronto: Canadian Scholars' Press.

—— 1978 [1845]. "Alienation and Social Classes." In *The Marx-Engels Reader*, edited by Robert C. Tucker, 133–135. New York: Norton.

Masjuan, Eduard. 2000. *La Ecología Humana en el Anarquismo Ibérico: Urbanismo Orgánico o Ecológico, Neomalthusianismo y Naturismo Social*. Barcelona: Icaria.

Matea Rosa, María L. 2013. "El Fondo de Titulación del Déficit del Sistema Eléctrico." *Boletín Económico de ICE* 3039:15–23.

Mayayo, Andreu. 1995. *De Pagesos a Ciutadans: 100 Anys de Sindicalisme i Cooperativisme Agraris a Catalunya (1893–1994)*. Catarroja (Spain): Afers.

McCrudden, Christopher, ed. 2013. *Understanding Human Dignity*. Oxford: Oxford University Press.

Melosi, Martin. 2006. "Energy Transitions in Historical Perspective." In *Energy and Culture: Perspectives on the Power to Work*, edited by Brendan Dooley, 3–18. Aldershot (UK): Ashgate.

Mestre, María Carmen. 1977. "Las Empresas Electricas durante la Crisis Energetica." *Investigaciones Económicas* 3:143–174.

Midnight Notes Collective. 1992 [1979]. "Strange Victories: The Anti-Nuclear Movement in the US and Europe." In *Midnight Oil: Work, Energy, War, 1973–1992*, edited by Midnight Notes Collective, 193–214. Brooklyn, NY: Autonomedia.

Mitchell, Catherine. 2010. *The Political Economy of Sustainable Energy*. Basingstoke (UK): Palgrave Macmillan.

Mitchell, Timothy. 2011. *Carbon Democracy: Political Power in the Age of Oil*. London: Verso.

—— 1990. "Everyday Metaphors of Power." *Theory and Society* 19(5):545–577.

Moncada, Jesús. 1995 [1988]. *The Towpath*. New York: Harpercollins.

Moore, Barrington. 1966. *Social Origins of Dictatorship and Democracy: Lord and Peasant in the Making of the Modern World*. Boston: Beacon Press.

Moore, Jason W. 2015. *Capitalism in the Web of Life: Ecology and the Accumulation of Capital*. London: Verso.

Moragues-Faus, Ana. 2014. How is Agriculture Reproduced? Unfolding Farmers' Interdependencies in Small-Scale Mediterranean Olive Oil Production. *Journal of Rural Studies* 34:139–151.

Morris, Brian. 2014. *Anthropology, Ecology, and Anarchism*. Oakland, CA: PM Press.

Mumford, Lewis. 2010 [1934]. *Technics and Civilization*. Chicago: University of Chicago Press.

Muñoz, Juan, Santiago Roldán, and Ángel Serrano. 1979. "The Growing Dependence of Spanish Industrialization on Foreign Investment." In *Underdeveloped Europe: Studies in Core-Periphery Relations*, edited by Dudley Seers et al., 161–176. Atlantic Highlands, NJ: Humanities Press.

Muro, Ignacio. 1994. "FECSA y la Reordenación del Sector Eléctrico, 1975–1990." In *Las Tres Chimeneas*, Volume 3, edited by Horacio Capel, 77–119. Barcelona: FECSA.

Nadai, Alain, Werner Krauss, Ana Afonso, Dorle Dracklé, and Oliver Hinkelbein. 2010. "El Paisaje y la Transición Energética: Comparando el Surgimiento de Paisajes de Energía Eólica en Francia, Alemania y Portugal." *Nimbus* 25/26:155–173.

Nader, Laura. 2013. *Culture and Dignity: Dialogues between the Middle East and the West.* Oxford: Wiley Blackwell.

———, ed. 2010. *The Energy Reader.* Oxford: Wiley Blackwell.

Nader, Laura and Stephen Beckerman. 1978. "Energy as it Relates to the Quality and Style of Life." *Annual Review of Energy* 3:1–28.

Naredo, José M. 2009. "Economía y poder. Megaproyectos, Recalificaciones y Contratas." In *Economía, Poder y Megaproyectos*, edited by Federico Aguilera and José M. Naredo, 19–52. Taro de Tahíche, Lanzarote (Spain): Fundación César Manrique.

——— 2001. *Por una Oposición que se Oponga.* Barcelona: Anagrama.

——— 1986. "La Agricultura en el Desarrollo Económico." In *Historia Agraria de la España Contemporánea III: 1900–1960*, edited by Ramon Garrabou et al., 455–498. Barcelona: Crítica.

Narotzky, Susana. 2016a. "Between Inequality and Injustice: Dignity as a Motive for Mobilization during the Crisis." *History and Anthropology* 27(1):74–92.

——— 2016b. "On Waging the Ideological War: Against the Hegemony of Form." *Anthropological Theory* 16(2/3):263–284.

——— 2016c. "Where Have All the Peasants Gone?" *Annual Review of Anthropology* 45:301–318.

——— 2007. "'A Cargo del Futuro'—Between History and Memory: An Account of the Fratricidal Conflict during Revolution and War in Spain (1936–39)." *Critique of Anthropology* 27(4):411–429.

——— 1997. *New Directions in Economic Anthropology.* London: Pluto Press.

Narotzky, Susana and Gavin Smith. 2005. *Immediate Struggles: People, Power and Place in Rural Spain.* Berkeley: University of California Press.

Nel·lo, Oriol. 1991. "Les Teories sobre l'Ordenament del Territori a Catalunya: Els Antecedents." *Papers: Regió Metropolitana de Barcelona* 6:77–90.

Nel·lo, Oriol, ed. 2003. *Aquí no! Els Conflictes Territorials a Catalunya.* Barcelona: Empúries.

Nichols, Robert. 2015. "Disaggregating Primitive Accumulation." *Radical Philosophy* 194:18–28.

Núñez, Gregorio. 1995. "Empresas de Producción y Distribución de Electricidad en España (1878–1953)." *Revista de Historia Industrial* 7:39–79.

Nye, David E. 1999. *Consuming Power: A Social History of American Energies.* Cambridge, MA: MIT Press.

Oceransky, Sergio. 2010. "Fighting the Enclosure of Wind: Indigenous Resistance to the Privatization of the Wind Resource in Southern Mexico." In *Sparking a Worldwide Energy Revolution*, edited by Kolya Abramsky, 505–522. London: AK Press.

O'Connor, James. 1998. *Natural Causes: Essays in Ecological Marxism*. New York: Guilford Press.

Pagès, Pelai. 2004. "La Fatarella: Una Insurrecció Pagesa a la Reraguarda Catalana durant la Guerra Civil." *Estudis d'Història Agrària* 17:659–674.

Palomera, Jaime. 2015. *The Political Economy of Spain: A Brief History (1939-2014)*. Working paper for the ERC Greco Project. Department of Anthropology, Universitat de Barcelona, 51 pages.

——— 2014. "How Did Finance Capital Infiltrate the World of the Urban Poor? Homeownership and Social Fragmentation in a Spanish Neighborhood." *International Journal of Urban and Regional Research* 38(1):218–235.

Palomera, Jaime and Theodora, Vetta. 2016. "Moral Economy: Rethinking a Radical Concept." *Anthropological Theory* 16(4):413–432.

Pasqualetti, Martin J. 2011. Opposing Wind Energy Landscapes: A Search for Common Cause." *Annals of the Association of American Geographers* 101(4):907–917.

Peix, Andreu. 1999. *25 Anys de la Unió de Pagesos (1974–1999)*. Lleida (Spain): Pagès Editors.

Perelman, Michael. 2000. *The Invention of Capitalism: Classical Political Economy and the Secret History of Primitive Accumulation*. Durham, NC: Duke University Press.

Pillado, Raúl. 1979. "La Manipulación de la Opinión Pública a través del Sistema Informativo." *Cuadernos del Ruedo Ibérico* 63–66:307–324.

Pinilla Navarro, Vicente, ed. 2008. *Gestión y Usos del Agua en la Cuenca del Ebro en el Siglo XX*. Zaragoza (Spain): Prensas Universitarias de Zaragoza.

Planas, Jordi and Samuel Garrido. 2006. "Sindicalisme, Cooperativisme i Conflictivitat Agrària en el Primer Terç del Segle XX." In *Història Agrària dels Països Catalans IV: Segles XIX–XX*, edited by Ramon Garrabou, 555–580. Barcelona: Fundació Catalana per a la Recerca.

Ploeg, Jan D. van der. 2013. *Peasants and the Art of Farming: A Chayanovian Manifesto*. Halifax, NS: Fernwood Publishing.

——— 2009. *The New Peasantries: Struggles for Autonomy and Sustainability in an Era of Empire and Globalization*. Oxon (UK): Earthscan.

Podobnik, Bruce. 2006. *Global Energy Shifts: Fostering Sustainability in a Turbulent Age*. Philadelphia: Temple University Press.

Polanyi, Karl. 2001 [1944]. *The Great Transformation: The Political and Economic Origins of Our Time*. Boston: Beacon.

Pomeranz, Kenneth. 2000. *The Great Divergence: China, Europe, and the Making of the Modern World Economy*. Princeton, NJ: Princeton University Press.

Pont, Josep, ed. 2002. *La Lluita per l'Ebre: El Moviment Social Contra el Pla Hidrològic Nacional*. Barcelona: Mediterrània.

Presas, Oleguer. 2014. *La Fatarella: Transformació d'un Territori. La Implantació d'una Central Eòlica en un Territori Rural*. Unpublished report, sponsored by Fundació El Solà, 90 pages.

Prieto, Pedro and Charles A. S. Hall. 2013. *Spain's Photovoltaic Revolution: The Energy Return on Investment*. London: Springer.

Powell, Dana E. and Dáilan J. Long. 2010. "Landscapes of Power: Renewable Energy Activism in Diné Bikéyah." In *Indians & Energy: Exploitation and Opportunity in the American Southwest*, edited by Sherry L. Smith and Brian Frehner, 231–262. Santa Fe, NM: SAR Press.

Puig i Boix, Josep. 2014. "Wind Energy in Spain." In *Wind Power for the World: International Reviews and Developments*, edited by Preben Maegaard et al., 473–492. Singapore: Pan Stanford Publishing.

———— 2009. "Renewable Regions: Life after Fossil Fuels in Spain." In *100% Renewables: Energy Autonomy and Action*, edited by Peter Droege, 187–204. London: Earthscan.

Rebés, Xavier. 2003. *La Pedra en Sec a la Fatarella*. Calaceit (Spain): Fundació el Solà.

Rebull, Joan. 1979. *La Protesta Nuclear a Catalunya: Una Opció Energètica Contestada*. Barcelona: Fundació Roca i Galès.

Recasens Llort, Josep. 2005. *La Repressió Franquista a la Terra Alta (1938–1945)*. Horta de Sant Joan (Spain): Ecomuseu dels Ports.

Redclift, Michael. 2005. "Sustainable Development (1987–2005): An Oxymoron Comes of Age." *Sustainable Development* 13(4):212–227.

Regueiro Ferreira, Rosa M. 2011. *El Negocio Eólico: La Realidad del Empleo, Promotores y Terrenos Eólicos*. Madrid: Catarata.

Reyna, Stephen P. and Andrea Behrends. 2011. "The Crazy Curse and Crude Domination." In *Crude Domination: An Anthropology of Oil*, edited by Andrea Behrends et al., 3–29. New York and Oxford: Berghahn.

Ribeiro, Gustavo L. 1994. *Transnational Capitalism and Hydropolitics in Argentina: The Yaciretá High Dam*. Gainesville, FL: University Press of Florida.

Ribot, Jesse C. and Nancy L. Peluso. 2003. "A Theory of Access." *Rural Sociology* 68(2):153–181.

Riley, Stephen. 2013. "The Function of Dignity." *Amsterdam Law Forum* 5(2):90–106.

Robbins, Paul. 2012. *Political Ecology*, 2nd ed. Oxford: Wiley-Blackwell.

Roig Amat, Barto. 1970. *Orígenes de la Barcelona Traction (Conversaciones con Carlos E. Montañés)*. Pamplona (Spain): Ediciones Universidad de Navarra.

Rosen, Michael. 2012. *Dignity: Its History and Meaning*. Cambridge, MA: Harvard University Press.

Ross, Kristin. 2015. *Communal Luxury: The Political Imaginary of the Paris Commune*. London: Verso.

Roy, Arundhati. 1999. *The Cost of Living*. New York: Modern Library.

Saladié, Sergi. 2013. *Impacte Econòmic de l'Energia Eòlica a Escala Local a Catalunya: Estudi Comparatiu*. Unpublished report, Universitat Rovira i Virgili (Tarragona, Spain), 88 pages.

———— 2011a. "Els Conflictes Territorials del Sistema Elèctric a Catalunya." *Treballs de la Societat Catalana de Geografia* 71/72:201–221.

———— 2011b. "Iniciativas de Gestión y Ordenación para la Sostenibilidad del Paisaje en la Comarca del Priorat (Cataluña)." *Actes Colloque International Paysages de la Vie Quotidienne: Regards Croisés entre la Recherche et l'Action*, Cemagref, Sciences, Eaux & Territoires, Perpignan-Girona.

Sánchez Cervelló, Josep. 2001. *Conflicte i Violència a l'Ebre: De Napoleó a Franco*. Barcelona: Flor del Vent.

Sánchez Cervelló, Josep (coord). 2003. *Maquis: El puño que Golpeó al Franquismo*. Barcelona: Flor del Viento.

Sánchez-Herrero, Mario. 2014. "Las Renovables Vistas por los Señores de la Energía." In *Alta Tensión*, edited by JV Barcia Magaz and Cote Romero, 97–102. Barcelona: Icaria.

Scheer, Hermann. 2006. *Energy Autonomy: The Economic, Social, and Technological Case for Renewable Energy*. London: Earthscan.

Schneider, François, Giorgios Kallis, and Joan Martínez-Alier. 2010. "Crisis or Opportunity? Economic Degrowth for Social Equity and Ecological Sustainability." *Journal of Cleaner Production* 18:511–518.

Scoones, Ian. 2015. *Sustainable Livelihoods and Rural Development*. Black Point, NS: Fernwood Publishing.

Scott, James C. 1985. *Weapons of the Week: Everyday Forms of Peasant Resistance*. New Haven, CT: Yale University Press.

—— 1976. *The Moral Economy of the Peasant: Rebellion and Subsistence in Southeast Asia*. New Haven, CT: Yale University Press.

Scott, Rebecca J. 2013. "Dignité/Dignidade: Organizing Threats to Dignity in Societies after Slavery." In *Understanding Human Dignity*, edited by Christopher McCrudden, 61–77. Oxford: Oxford University Press.

Sempere, Joaquim. 2008. "Energia Eòlica i Territori: Arguments i Protestes. El Cas dels Projectes Eòlics de la Terra Alta." In *Societat Catalana 2008*, edited by Teresa Montagut, 139–156. Barcelona: Associació Catalana de Sociologia.

Serrano, Ángel and Juan Muñoz. 1979. "La Configuración del Sector Eléctrico y el Negocio de la Construcción de las Centrales Nucleares." *Cuadernos del Ruedo Ibérico* 63–66:127–267.

Sevilla Guzmán, Eduardo. 1979. *La Evolución del Campesinado en España: Elementos para una Sociología Política del Campesinado*. Barcelona: Península.

Sevilla Guzmán, Eduardo, Miguel Doñate, Raúl Márquez, and Pablo Romero. 2008. "Conversando con Eduardo Sevilla Guzmán." *(Con)textos: Revista d'Antropologia Social* 2:5–17.

Shanin, Teodor. 1986. "Chayanov's Message: Illuminations, Miscomprehensions, and the Contemporary 'Development Theory'." In *The Theory of Peasant Economy*, edited by Daniel Thorner et al., 1–24. Madison, WI: University of Wisconsin Press.

Shever, Elana. 2012. *Oil and Reform: Oil and Neoliberalism in Argentina*. Stanford, CA: Stanford University Press.

Sider, Gerald and Gavin Smith, eds. 1997. *Between History and Histories: The Making of Silences and Commemorations*. Toronto: University of Toronto Press.

Sieferle, Rolf P. 2001 [1982]. *The Subterranean Forest: Energy Systems and the Industrial Revolution*. Cambridge: White Horse Press.

Simoni, Encarna and Renato Simoni. 1984. *Cretas: La Colectivizaciín de un Pueblo Aragonés durante la Guerra Civil Española (1936–37)*. Alcañiz (Spain): Centro de Estudios Bajoaragoneses.

Smil, Vaclav. 2010. *Energy Transitions: History, Requirements, Prospects*. Santa Barbara, CA: ABC-CLIO.

—— 2008. *Energy in Nature and Society: General Energetics of Complex Systems*, 2nd ed. Cambridge, MA: MIT Press.

Smith, Gavin. 2016. "Against Social Democratic Angst about Revolution." *Dialectical Anthropology* 40(3):221–239.

—— 2011. "Selective Hegemony and Beyond—Populations with "No Productive Function": A Framework for Enquiry. *Identities* 18(1):2–38.

Smith, Neil. 2008 [1984]. *Uneven Development: Nature, Capital, and the Production of Space*. Athens, GA: University of Georgia Press.

Solé Sabaté, Josep M. 2003. *La Repressió Franquista a Catalunya (1938–1953)*. Barcelona: Edicions 62.

Soler, Carles and Fernando Fernández. 2015. *Estructura de la Propiedad de Tierras en España: Concentración y Acaparamiento*. Bilbao (Spain): Mundubat.

Soronellas, Montserrat. 2006. *Pagesos en un Món de Canvis: Famílies i Associacions Agràries*. Tarragona (Spain): Universitat Rovira i Virgili.

Sorribes, Jesús and Josep-Joan Grau Folch. 1989. *La Ribera d'Ebre: Transformacions Socio-Econòmiques i Perspectives de Futur*. Barcelona: Caixa de Catalunya.

Strathern, Marilyn. 2004. *Partial Connections*. Updated Edition. Walnut Creek, CA: AltaMira.

Strauss, Sarah, Stephanie Rupp, and Thomas Love, eds. 2013. *Cultures of Energy: Power, Practices, and Technologies*. Walnut Creek, CA: Left Coast Press.

Sudrià, Carles. 1987. "Un Factor Determinante: La Energía." In *La Economía Española en el Siglo XX: Una Perspectiva Histórica*, edited by Jordi Nadal et al., 313–363. Barcelona: Ariel.

Svampa, Maristella. 2013. "'Consenso de los Commodities' y Lenguajes de Valoración en América Latina." *Nueva Sociedad* 244:30–46.

Swyngedouw, Erik. 2015. *Liquid Power: Water and Contested Modernities in Spain, 1898–2010*. Cambridge, MA: MIT Press.

———2007. "Technonatural Revolutions: The Scalar Politics of Franco's Hydro-Social Dream for Spain, 1939–1975. *Transactions, Institute of British Geographers* NS 32:9–28.

Szarka, Joseph. 2007. *Wind Power in Europe: Politics, Business, and Society*. Basingstoke (UK): Palgrave Macmillan.

TARA (Tecnologías Alternativas Radicales y Autogestionadas). 1980. *Energías Libres II*. Zaragoza (Spain): Ecotopia.

———1977. Special Issue: La Cara Oculta de la Energía. *Alfalfa: Crítica Ecológica y Alternativas*.

Tarroja, Àlex et al. 2003. "Terres de l'Ebre: Una Identitat i un Projecte de Futur." *Papers, Regió Metropolitana de Barcelona* 39:151–181.

Taussig, Michael. 1980. *The Devil and Commodity Fetishism in Latin America*. Chappel Hill, NC: University of North Carolina Press.

Tébar, Javier. 2006. "Guerra, Revolució i Contrarevolució al Camp." In *Història Agrària dels Països Catalans IV: Segles XIX–XX*, edited by Ramon Garrabou, 581–602. Barcelona: Fundació Catalana per a la Recerca.

Termes, Josep. 2008. *Misèria contra Pobresa: Els Fets de la Fatarella del Gener de 1937*. Barcelona: Editorial Afers.

Terradas, Ignasi. 1984. *El Món Històric de les Masies*. Barcelona: Curial.

———1979. *Les Colònies Industrials: Un Estudi entorn del Cas de l'Ametlla de Merola*. Barcelona: Laia.

Thompson, Edward P. 1971. "The Moral Economy of the English Crowd in the Eighteenth Century." *Past & Present* 50:76–136.

Toke, Dave. 2002. "Wind Power in UK and Denmark: Can Rational Choice Help Explain Different Outcomes?" *Environmental Politics* 11(4):83–100.

Tomàs, Manel. 2002. "Prefaci." In *La Lluita per l'Ebre: El Moviment Social contra el Pla Hidrològic Nacional*, edited by Josep Pont, 7–10. Barcelona: Mediterrània.

Trawick, Paul and Alf Hornborg. 2015. "Revisiting the Image of Limited Good: On Sustainability, Thermodynamics, and the Illusion of Creating Wealth." *Current Anthropology* 56(1):1–27.

Tsing. Anna L. 2015. *The Mushroom at the End of the World: On the Possibility of Life in Capitalist Ruins*. Princeton, NJ: Princeton University Press.

Trotsky, Leon. 1906. *Results and Prospects*. Available at: https://www.marxists.org/archive/trotsky/1931/tpr/rp-index.htm.

Urry, John. 2014. "The Problem of Energy." *Theory, Culture & Society* 31(5):3–20.

Vaccaro, Ismael and Oriol Beltran. 2008. "The New Pyrenees: Contemporary Conflicts around Patrimony, Resources and Urbanization." *Journal of the Society for the Anthropology of Europe* 8(2):4–15.

Varillas, Benigno and Humberto da Cruz. 1981. *Para una Historia del Movimiento Ecologista*. Madrid: Miraguano.

Vernet, Roser. 2008. "La Implantació Eòlica al Priorat." In *La Cultura del No: El Conflicte Ambiental i Territorial a Catalunya*, edited by Antoni Ferran and Carme Casas, 98–103. Vic (Spain): Eumo.

Vilar, Pierre. 1964–1968. *Catalunya dins l'Espanya Moderna: Recerques sobre els Fonaments Econòmics de les Estructures Nacionals*. 4 Volumes. Barcelona: Edicions 62.

Wainwright, Joel. 2010. "On Gramsci's 'Conception of the World.'" *Transactions of the Institute of British Geographers* NS 35:507–521.

Waldron, Jeremy, Wai Chee Dimock, Don Herzog and Micheal Rosen. 2015. *Dignity, Rank, & Rights*. Oxford: Oxford University Press.

Watts, Michael. 2004. "Resource Curse? Governmentality, Oil and Power in the Niger Delta, Nigeria." *Geopolitcs* 9(1):50–80.

Werner, Marion. 2016. *Global Displacements: The Making of Uneven Development in the Caribbean*. Oxford: Wiley-Blackwell.

White, Leslie A. 1943. "Energy and the Evolution of Culture." *American Anthropologist* 54(3):335–356.

Whitehead, Judith. 2010. *Development and Dispossession in the Narmada Valley*. Delhi: Pearson.

Wilk, Richard. 2007. "The Extractive Economy: An Early Phase of the Globalization of Diet, and its Environmental Consequences." In *Rethinking Environmental History: World-System History and Global Environmental Change*, edited by Alf Hornborg et al, 179–198. Lanham, MD: AltaMira.

Williams, Raymond. 1980. "Ideas of Nature." In *Problems in Materialism and Culture*, 67–85. London: Verso.

———1977. *Marxism and Literature*. Oxford: Blackwell.

———1973. *The Country and the City*. Oxford: Oxford University Press.

Wolf, Eric R. 2001. *Pathways of Power: Building an Anthropology of the Modern World*. Berkeley, CA: University of California Press.

———1969. *Peasant Wars of the Twentieth Century*. New York: Harper & Row.

Wolsink, Maarten. 2007. "Wind Power Implementation: The Nature of Public Attitudes: Equity and Fairness Instead of 'Backyard Motives'." *Renewable and Sustainable Energy Reviews* 11(6):1188–1207.

Wood, Ellen Meiksins. 2015 [1989]. *Peasant-Citizen and Slave: The Foundations of Athenian Democracy*. London: Verso.

—— 2002. *The Origin of Capitalism: A Longer View*. London: Verso.

Wright, Melissa W. 2006. *Disposable Women and Other Myths of Global Capitalism*. New York: Routledge.

York, Richard. 2012. "Do Alternative Energy Sources Displace Fossil Fuels?" *Nature, Climate, Change* 2:441–443.

Zizek, Slavoj. 2006. *How to Read Lacan*. Cambridge: Granta Books.

Zografos, Christos and Sergi Saladié. 2012. "La Ecología Política de Conflictos sobre Energía Eólica: Un Estudio de Caso de Cataluña." *Documents d'Anàlisi Geogràfica* 58(1):177–192.

Zografos, Christos and Joan Martínez-Alier. 2009. "The Politics of Landscape Value: A Case Study of Wind Farm Conflict in Rural Catalonia." *Environment and Planning A* 41:1726–1744.

Zonabend, Françoise. 1989. *The Nuclear Peninsula*. Cambridge: Cambridge University Press.

—— 1980. "Le nom de Personne." *L'Homme* 20(4):7–23.

Index

JAUME FRANQUESA is Assistant Professor in the Department of Anthropology of the University at Buffalo, The State University of New York. He is the author of *Urbanismo neoliberal, negocio inmobiliario y vida vecinal: El caso de Palma.*

www.ingramcontent.com/pod-product-compliance
Lightning Source LLC
Chambersburg PA
CBHW071846270326
41929CB00013B/2124